THE ROMAN
WAR
MACHINE

The east Roman army and navy in about AD 400, from a sixteenth-century drawing of the Column of Arcadius

THE ROMAN
WAR
MACHINE

John Peddie

SUTTON PUBLISHING

First published in 1994 by
Sutton Publishing Limited · Phoenix Mill · Thrupp
Stroud · Gloucestershire

Paperback edition first published in 1996

A catalogue record of this book is available from the British Library

ISBN 0-7509-1023-2

Cover illustration: The Antonine Wall in Scotland, a 'distance slab' recording a
sector constructed by a named unit, the Second Legion *Augusta. (Trustees of the
National Museum of Scotland)*

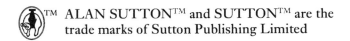

ᵀᴹ ALAN SUTTONᵀᴹ and SUTTONᵀᴹ are the
trade marks of Sutton Publishing Limited

Typeset in 11/12 Ehrhardt
Typesetting and origination by
Sutton Publishing Limited
Printed in Great Britain by
Hartnolls, Bodmin, Cornwall

Contents

They are quick to act as they are slow
to give way, and never was there an
engagement in which they were worsted by
numbers, tactical skill or unfavourable
ground – or even by fortune, which is less
within their grasp than victory.

Josephus (AD 37–93)

List of Illustrations

The author and publishers gratefully acknowledge the following for their permission to reproduce illustrations: B.T. Batsford/RCHM (46); Bodleian Library (3); The Carson Clark Gallery, Edinburgh (2); Colchester Museums (47); Corbridge Roman Museums (48); Egypt Exploration Society, London (40); Ermin Street Guard (6, 11, 21, 36); Paul Lewis Isemonger (10, 15, 22, 28); Oscar and Peter Johnson Ltd (34); Frank Lepper and Sheppard Frere, authors of *Trajan's Column* (Alan Sutton Publishing, 1988) (8, 19, 20, 26, 29, 35, 42); The Mansell Collection (4, 5, 7, 13, 33, 41, 49, 54, 59); Museo Lapidario Maffeino, Verona (Umberto Tomba) (17); Oxford University Press (39); Rheinisches Landesmuseum Trier (18, 23); Margaret Rylatt (27); Society of Antiquaries/RCHM (38); Trinity College Library, Cambridge (1).

Introduction

Early historians have provided us with a wealth of literature on Roman military matters. Some, of whom perhaps Onasander is the most illuminating, have bequeathed us treatises on the qualities required for Roman generalship. Many, such as Julius Caesar, Josephus, Livy and Ammianus Marcellinus, writing from practical experience, have left us colourful accounts of battles in which military tactics of the day are well described. Frontinus has condensed extracts from these works into a book of *Stratagems*, some bizarre but many remarkable in their similarity to modern-day tactics – 'Go for the high ground!' has long been the cry of the infantry, it appears. The works of other authors, among whom number Aenas and Vegetius, have consolidated contemporary military practice into manuals and textbooks, and have assembled for us technical accounts of ancient battle drills and tactical formations, as well as interesting detail of day-to-day administration and the occasional nugget of strategical thinking; but, almost invariably, much of what we would wish to know about their command techniques, staff planning arrangements and logistical techniques is omitted.

As a consequence, we are frequently left to ponder how Roman generals achieved so much with such seemingly scant resources and we are occasionally tempted to draw wrongful conclusions. 'The truth is', one military historian has written, 'Julius Caesar was not an organiser; careful preparations – adequate supplies, sufficient fighting forces and many other requirements needed to assure the success of a campaign – were either distasteful to him or lost to sight by reliance on his genius to solve all difficulties.'[1] This is an accusation which reads uneasily when set against a list of Caesar's military successes, for it was surely this genius, with the great man's mastery of manoeuvre and his ability to surprise and destabilize his opponents, which enabled Caesar to emerge victorious from his many campaigns, particularly as conqueror of Gaul. Sound administration is a natural prerequisite for the success of any operation. On the other hand, the nice balance between planned administration and operational arrangements, which must inevitably vary according to the military situation, is something which can only be determined by the commanding general; and his degree of success in doing this establishes his reputation.

Nevertheless, if it were true that Julius Caesar placed his supply arrangements at risk in order to achieve his objectives, he would not have been the first general to do so – he was certainly not the last. Indeed, he may be thought to have had much in common with the spirited commander of the German Afrika Korps,

Field Marshal Erwin Rommel, who likewise earned a reputation as an improviser with a tendency to impetuosity. Rommel's dashing leadership, despite his scarce resources and the initial orders he received from his superiors to limit his main activity to the defence of Tripolitania, brought him close to total victory against the British in 1942, when he drove his army headlong along the North African coast in an effort to seize the Nile basin, together with the strategically important Suez Canal.[2] He failed only by the narrowest of margins. In Burma in 1945 Lieutenant General Sir William Slim, commanding the XIVth Army, took a similar chance. He concluded that, if he were to capture Rangoon before the coming of the monsoon, he would have to commit all his supply vehicles to the advance and, as a consequence, impose half-rations on his soldiers. He wrote,

> It was very plain to me – and if it had not been, plenty of people were willing to enlighten me – that this dash for Rangoon by a mechanised force, confined to one road, thrusting against time through superior numbers, was a most hazardous and possibly rather un-British operation. I knew the risks and the penalties of failure but, as I checked over the final plans, I was ready to accept them.[3]

Sometimes, therefore, it is militarily necessary to take calculated risks, the nub being that the commander should be sufficiently competent to recognize when the culminating point has been reached and that the moment has arrived to revert to the defensive in order to rebuild his fighting power. This is a recognized tenet of military philosophy. We may assume that Julius Caesar, a natural fighting soldier, would have been prepared to take such risks as much as any other general of his, or later, generations; but a charge that careful preparations were distasteful to him can surely only be levelled if it can be shown that he had an inadequate back-up organization to support him. There is no clear evidence of this and it is a matter to which we shall return in the chapters which follow.

It is a truism to say that the supply arrangements of an army vary according to the fruitfulness of its theatre of operations. In Gaul (54–51 BC), Caesar was fortunate to be campaigning in a prosperous grain-producing country, with powerful allies nearby, albeit his enemies frequently subjected him to a ruthless 'scorched earth' policy. Four years later, during the opening weeks of his North African campaign, he faced a different situation, for here he was compelled to send a considerable distance, 'to Sardinia and the other neighbouring provinces', for reinforcements, supplies and corn.[4] In the following century, during the 70s, when the emperor Vespasian was on the throne, foodstuffs, *materiel* and men were being despatched to Rome's army on the Armenian front, along a lengthy supply line, reaching from the Danube estuary to the Black Sea port of Trapezus and beyond. Such intricate arrangements could surely not have been contemplated, set in place and carried forward without the existence of an accomplished supply organization briefed with the task of establishing sources of provision, systems of transportation and the means of getting reinforcements and stores forward to the fighting forces in the field.

Such an organization implies the existence of a staff corps, of which Josephus[5] may have been hinting when he penned the words that, in Roman war, nothing is done without plan or on the spur of the moment; careful thought precedes action of any kind, and to the decisions reached all ranks must conform. . . . They regard success due to luck as less desirable than a planned but unsuccessful stroke, because victories that come of themselves tempt men to leave things to chance, but forethought, in spite of occasional failures, is good practice in avoiding the same mistake. . . . Unfortunate accidents that upset calculations have at least this comfort in them, that plans were properly laid.

Likewise, Vegetius,[6] in his remark that the Romans owed the conquest of their world to high standards of military training and discipline and 'the unwearied cultivation of the other arts of war', appears to confirm the presence of such a staff structure, with military schools for its training and education. Time and again, throughout the writings of the ancients, we encounter clues to the existence of such a body. Julius Caesar's briefing of his generals and staff at sea, immediately prior to his landing in Britain (together with their remarkable presence at that moment when they might otherwise have been expected to be with their formations); the huge concentrations of siege artillery at Alesia under Caesar and at Jerusalem under Titus, both of which would have demanded decentralized command if the weapons were to be flexibly employed; and the highly detailed and effective preparations for the Claudian invasion of Britain, with its sophisticated arrangements for the seaborne landing,[7] are just some examples which would have required a planning body of this nature. In effect, in an age when armies numbered scores of thousands of animals and men and their movement and manoeuvring, without the benefits of modern-day transport and communications, would have posed major problems of command and control, the presence of some such supervisory staff structure would have been essential.

Equally, there are other aspects of Roman war, discussed within the pages of this book, about which much information is lacking. The employment of animals, for example, used prolifically by the ancients both for cavalry and transportation, raises a host of questions, mainly logistical. When the huge numbers involved are set alongside the defensive and administrative demands of the marching-camp technique, which required an army on the march to entrench itself behind ramparts at the end of every day, the immediately apparent answers quickly become shrouded in doubt. What did they do with the horses? remains a cry frequently raised and not easily answered, when considered in the light of the competing needs of grazing areas, control and security from ambush.

There are three constant factors upon which military historians may depend when glancing into the distant past. First, armies, like men, cannot survive without essentials such as food and drink. Where forage and provisions have not been carefully provided, wrote Vegetius, whose treatise *Epitoma rei militaris* became the military bible of European leaders well into medieval times, the evil is without remedy. The movement of armies is therefore guided by the whereabouts of these important necessities, or by the lack of them. A second factor is the Earth's relatively rarely changing landscape, the configuration of which constantly dictates the scope of the tactical decisions a commander may adopt, limiting or broadening his choice. The third is the predictable thinking process of man himself, particularly

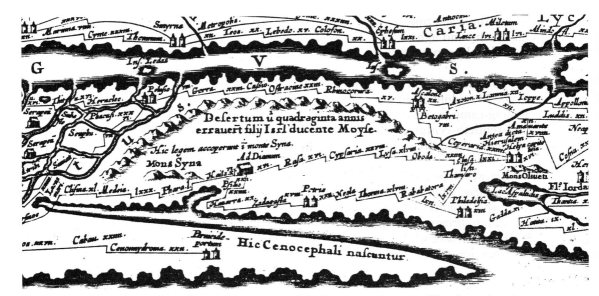

Conrad Peutinger's (1465–1547) road map of the Roman empire exemplifies the unchanging nature of the military world. In 1948 an Israeli armoured column from Jerusalem captured Eilat by using an old Roman road (see Appendix 1)

when educated within the disciplines of a military environment. There is a further, fourth, consideration, argued by Field Marshal Viscount Montgomery of Alamein,[8] namely that, even though weapons have become more powerful and the problems of the battlefield more complex, the art of war is fundamentally the same today as it was in the ancient world. If we bear these important guidelines in mind when searching for the answers which elude us, we will discover we are erecting signposts which point to areas where they may be hidden.

When thumbing through old reference books, I have repeatedly been reminded of Montgomery's words. In most modern battalion 'Standing Orders for War', for example, it is laid down that every man should have a hot meal before battle. This advice is also sensibly voiced by Onasander in his treatise, *The General*.[9] Here he writes that senior officers 'should not hesitate to order the first meal at sunrise, lest the enemy, by a surprise attack, force his men to fight while still hungry. . . .This matter should not be considered of slight importance nor should a general neglect to pay attention to it, for soldiers who have eaten moderately . . . are more vigorous in battle.' Similarly, any soldier who has experienced the problems of trying to eat while actively engaged in the field, will recognize the drill initiated by Philip of Macedonia in 200 BC, during battle with the Aetolians, when

he ordered the men to fetch water and to have their meal, in succession, the cavalry by squadrons, and the light infantry by companies, while he kept some on guard under arms, as he waited for the column of infantry, which moved more slowly under the weight of their equipment. On arrival they were ordered to plant their standards, pile their arms, and make a hasty meal, not more than

two or three at a time being sent from each company to fetch water; meanwhile
the cavalry and light infantry stood in formation, on the alert in case the enemy
should make a move.[10]

Livy's description of Paulus's systematic withdrawal of his troops from battle
with the Macedonians in 168 BC, to take up positions behind the ramparts of his
marching–camp, is yet another case in point.[11] It might have been quoted in any
good twentieth-century military training manual as an example of the care to be
taken under these circumstances:

> he first withdrew the *triarii* from the rear of the battle–line; then he brought back
> the *principes*, while the *hastati* maintained their position in the front of the line, in
> case the enemy should make a move; lastly, he gradually withdrew
> the *hastati*, bringing off the soldiers of one maniple at a time, beginning with the
> right wing. Thus, while the cavalry and light armed remained facing the enemy in
> the front line of battle, the infantry were withdrawn without commotion.

There is no shortage of such examples, as Vegetius's rhetorical question
demonstrates,[12] when he asks whether a man may be reckoned a good soldier 'who
through negligence allows his arms to suffer by dirt and rust?'. It arouses memories
of those countless nights of war, when weapons were produced, cleaned and lightly
oiled in preparation for the morrow.

The basic arts of soldiering have surely changed but little in two thousand years
and more.

A page from the Notitia Dignitatum *showing the insignia of the first secretary of the Chancellery. The* Notitia *is an invaluable source for the structure and organization of the Roman army*

CHAPTER ONE

Roman Generalship

'A clock is like an army,' I used to tell them. 'There's a main spring, that's the Army Commander, who makes it all go; then there are other springs, driving the wheels round, those are his generals. The wheels are officers and men. Some are big wheels, very important, they are the chief staff officers and the colonel sahibs. Other wheels are little ones, that do not look at all important. They are like you. Yet stop one of those little wheels and see what happens to the rest of the clock!'

Field Marshal Sir William Slim, *Defeat into Victory*[1]

The fundamental problems that have confronted armies and their commanders over the years – the need for a secure base; the sustenance of the army in the field; the provision of trained manpower and the creation of the right political climate – have always been factors of prime consideration. Nevertheless, the various ways in which they have been resolved differ greatly, and no more so than in the case of ancient Rome as it developed from a city-state.

The highest office in Republican Rome was vested in a pair of magistrates, consuls, who were elected annually by an assembly of the people and held power alternately. They were advised in their duties by the Senate, a body of ex-magistrates. The precise extent of this advice is not clear but, plainly, it could have acted as an inhibiting brake when urgent decision-making was required. The reason for joint command by consuls is obvious: each provided a check on the other and thus reduced the opportunity for either to seize power and retain authority, but it was a watchfulness which verged on impracticality. In time of war, each consul commanded the army (then standardized at a strength of four legions, plus allies), on alternate days. There was another complicating factor, particularly in the Republican era, which might have been thought to dilute efficiency still further, if, strangely, it had not so often proved to the contrary. All important magistracies, governorships and army commands were held by senators, generally without any military ability, their qualifications resting largely on their wealth and their circumstance of birth. Yet the positions to which they aspired were recognized as being of great import to the nation. 'For who can doubt', commented Vegetius, 'that military skill is superior to all other accomplishments, since through it our freedom and authority are preserved, our territory enhanced and our empire safeguarded.'[2]

The frailties of the command structure, and the problems created by this lack of continuity of leadership, were recognized by the Senate in two ways: firstly, on the outbreak of war, by the appointment of a supreme commander, a *dictator*, but, even then, for the restricted period of one campaigning season only; secondly, by the re-appointment of consuls at the completion of their tour of duty or, alternatively, by allowing consuls to retain their military status and carry forward their service in the rank of proconsul. It was a system which produced some extraordinary successes, but more because of its non-observance than because of strict adherence to it. Gaius Marius, to whom are attributed many of the major reforms which moved the Roman army forward to near-professionalism, held an unprecedented series of consulships in the decades at the end of the second and the beginning of the first centuries BC. He was contemptuous of the Senate but popular among the masses, in part because of his declared intention not to force men into military service against their will but largely because, in an age when Italy was under threat of invasion, he was an able commander with a bent for leadership and a proven success record in the field. To the people he was a military hero; to his men, to whose training and welfare he gave unflagging attention, he was 'a soldier's soldier', a hard but fair man, who shared their way of life and understood their needs. In battle, he had a reputation for courage and unhurried thoroughness.

Julius Caesar, nephew of Gaius Marius, one-time governor of Farther Spain (61–60 BC), conqueror of Gaul (58–50 BC), victor in the Civil War (49–46 BC), consul (49 BC) and *dictator* (46–44 BC), was another product of the system. In history as in life, his name provokes controversy. 'Why, man,' eulogized William Shakespeare, through the mouth of Cassius,

> . . . he doth bestride the narrow world
> Like a Colossus, and we petty men
> Walk under his huge legs and peep about
> To find ourselves dishonourable graves.
> *Julius Caesar*, I, ii, 134–7

Eminent historians have opined that 'no commander who ever lived showed greater military genius',[3] and have referred to him variously as 'the entire and perfect man'[4] and 'the greatest man of action who has ever lived',[5] but such views are by no means universal. Professor Hugh Last, writing in the *Cambridge Ancient History*,[6] while acknowledging his leadership qualities, made the controversial declaration that Caesar bequeathed to those who followed him 'no receipt for victory except one beyond their reach – to be Caesar . . . he made no single innovation in the technique of soldiering'. Colonel Theodore Dodge, the distinguished American military historian, held him to be 'the greatest man in antiquity',[7] but also remarked that, if the months were counted, 'more than half of Caesar's campaigns were consumed in extricating himself from the results of his own mistakes'.[8] Major General J.F.C. Fuller extended this criticism by asserting unequivocally that, even though these extrications may have been brilliant, Caesar was guilty of 'incontestably bad generalship':[9]

His defective system of supply frequently compelled him to change his area of operations to his disadvantage; his untrained, barbaric cavalry at times led him

Julius Caesar, conqueror of Gaul and dictator, 48–44 BC: '. . . he doth bestride the narrow world/ Like a Colossus'.

into critical situations; and when his enemy took to guerilla warfare, he could do nothing to respond to him. It is astonishing that a soldier of his outstanding intelligence . . . could have failed to realise how defective was his army organisation.

This is strong meat and, within these pages, we shall be examining this statement further and aspiring to contest the 'incontestable'.

The policy of extending consular service, which applied equally to dictatorships,[10] was one of vital importance for, so far as is known, the Romans had no military training institutions and no formal process of either testing or educating officers in staff duties and the problems of command. In the absence of such establishments, they looked instead to experience and it has been suggested[11] that, until the last years of the second century BC, candidates for office had to serve a minimum number of campaigning seasons in order to be eligible for consideration, a sort of staff attachment, which Pliny,[12] in the reign of Domitian (AD 81–96), described when he wrote that

> in olden times there was a custom by which we acquired knowledge not just by listening to our elders but also by watching their conduct. In this way we learnt what we had to do ourselves and what should be passed on to our juniors. So, men were immersed in military service at an early age and learnt how to give orders by obeying them, how to be a commander by following others.

There are, however, two very relevant factors which should not be overlooked when considering the Roman army's needs for officer material; firstly, the core of battlefield veterans to be found among the centurions, a hardy, hand-picked body of men of great dependability and courage, who provided the army with a broad leavening of experience and military wisdom; and, among the staff, the presence of permanent civil servants responsible for military planning, whose availability within the secretariat is rarely recognized but whose office, inevitably, must have provided an invaluable breeding ground for staff officers. Nor is it difficult to visualize that planning teams from this 'ministry for war' would have been deployed at moments of crisis to assist governors of provinces and commanding generals in their work.

The likely staff role of the secretariat in these matters is underlined by the presence at Boulogne, prior to the Roman invasion of Britain in AD 43, of Narcissus, private secretary to the emperor Claudius. He was a former slave, who had risen to being clerk in the civil service and was later to be appointed secretary-general and head of the state department. It was fortunate he was there, for the army, grumbling at being ordered to serve outside 'the inhabited world', had mutinied and refused either to embark or listen to their general, Aulus Plautius. Narcissus, as the personal representative of the emperor, was now called upon to address them and was greeted, good-humouredly, by the cry of '*Io Saturnalia*', a reference to the fact that, at the feast of Saturn, slaves were permitted to appear wearing the clothes of their masters. In my book *Invasion*, I expressed the view that he had arrived at the coast to make arrangements for Claudius's impending arrival and departure to Britain; but this is likely to have been only part of the story. It is equally probable that he was present primarily as the emperor's chief staff officer, with the purpose of coordinating the plans which he, with his team from the secretariat, had outlined in Rome and finalized in Boulogne for the embarkation and departure of the task force and for the implementation of future supply arrangements. With a thousand ships to launch across the Channel, these plans would have been considerable and intricate, calling for a staff empowered to make instant decisions.

In the imperial period, the emperor was commander-in-chief and, nominally at least, led his army on to the field of battle or, in his absence, designated a member of his family to take his place. In the event of a war outside Italy, then it was customary either for a consul or ex-consul of suitable seniority and experience to be nominated to command the campaign or, alternatively, for its conduct to be passed to the governor of the territory or province concerned.

The sequence of the build-up to the invasion of Britain provides an example of this behaviour. The emperor Claudius selected Aulus Plautius, at that moment governor of Pannonia, to lead the expedition. Plautius was a kinsman of his first wife but it is more likely that his choice was influenced by the fact that the general had considerable experience of waterborne operations on the River Danube, working in close cooperation with the Roman navy. The task now confronting him posed similar demands. Plautius at once assumed command of the task force assembling at Boulogne on behalf of his commander-in-chief. When all was ready, he placed himself at the head of the invasion force, crossed the Channel and, when he felt assured that Colchester was ripe to fall into his hands, in accordance with his instructions, he sent for Claudius.[13] Upon the emperor's arrival from Rome, he handed command of the army to him. He re-assumed it once again, sixteen days later, when his master returned to Gaul.

This, then, was broadly the structure of command at the top, but what particular skills might have been sought of their generals by the Roman people? There were a multitude of views, as might be expected. Cicero, soldier and distinguished orator, opined that a good commander had to be a man of proven reputation, courage and good fortune, with a grasp of military science.[14] Onasander, in a treatise on generalship dedicated to Quintus Veranius, consul in AD 53, took a slightly divergent view. He underlined the necessity for common sense and strength of character but saw no great need for a detailed knowledge of technical matters: rather should a general take advice from men of experience. He should at all times lead by personal example but not by 'fighting in battle', a philosophy which would not have been supported by Julius Caesar. It is likely, however, that Onasander was suggesting that a commander should not get caught up in the cut and thrust of the fighting and thus find himself distracted from the duties of command. Scipio was probably indicating something similar when, accused of lacking aggression, he answered 'My mother bore me a general, not a warrior.'[15] Agricola, according to Tacitus,[16] took the field in person as soon as the campaigning season opened:

> He was present everywhere on the march, praising good discipline and keeping stragglers up to the mark. He himself chose sites for camps and reconnoitred estuaries and forests; all the time he gave the enemy no rest and constantly launched plundering raids.

Velleius,[17] writing of his hero Tiberius, stressed the virtue of caution; the safety of the army was paramount. Victory should not be sought through the sacrifice of Roman troops; a general should make his own decisions and never place his own reputation before common sense.

From the above paragraph, two things emerge: first, no one definition exists which can encompass the qualities of a general, particularly if he lies in the class defined by Montgomery as the *grand chef*; and, second, many qualities that today might be considered essential for inclusion have either been underemphasized or omitted – but need not necessarily have been unpractised.

Generals, whatever their generation, are expected to win wars and their reputations are gained almost as much by the extent of their successes as by the character of their campaigns. The basic ingredients that affect the conduct of a campaign are available to every commander; the phases of war, the nature of manoeuvre and the guiding principles (of which surprise, concentration of force, high morale, maintenance of the aim and sound administration number but a few) are unchanging and constant constituents. The greatest single winning factor is the spirit of the warrior himself. Thus to suggest, like Professor Hugh Last, that Caesar bequeathed no recipe for victory to those who followed after him, is to express an exaggerated confidence in the ability of commanders to conjure bright new methods of war from their knapsacks. Success comes from the artful degree to which all the ingredients listed above are, in the first place, appreciated and then handled and brought together.

This is a truism of which Julius Caesar possessed a natural understanding. He was fully conscious of the limitations of his tactical arsenal and squeezed the maximum benefits from it. He was the master of the Roman technique of the marching-camp (see Chapter 3) and of siege warfare, as he proved during his investment of Alesia during the Gallic rebellion. His chief weapon was the element of surprise, coupled with the scale, timing and speed of an attack. 'He joined battle', wrote Suetonius, 'not only after planning his movements in advance but on a sudden opportunity, often immediately at the end of a march, and sometimes in the foulest weather, when one would least expect him to make a move.'[18] There are numerous examples of the manner in which he achieved this: in 55 BC, in a move to discipline the Germans, the speed of his advance threw his enemy 'into sudden panic; they had no time to think what to do or to arm themselves'; in 52 BC, when operating against Vercingetorix in the Loire Valley, 'the shock of [his] appearance unnerved the enemy and the crossing was effected without loss': again, a year later, when dealing with a revolt by the Bituriges,[19]

Caesar's sudden march found them unprepared and scattered, tilling the fields without any thought of danger, and naturally they were caught by cavalry before they could take refuge in the town. They had not even the usual warning of a hostile invasion – the sight of burned buildings; for Caesar had commanded that nothing should be set on fire in order to avoid giving the alarm, and to save the corn and hay, which he would need if he decided to advance far.

In order to practise this philosophy to its utmost, the ability to strike with lightning suddenness was essential. In an age when the speed of advance was constrained to that of the supply train which moved with the infantry column, he frequently gambled by leaving his heavy baggage behind and travelled light, each man carrying operational rations and the minimum of equipment. Almost invariably, it profited him to do so. Occasionally, as at Ruspina, when he was said to be in a 'ferment of impatient expectancy'[20] awaiting the arrival of reinforcements and supplies, after having landed with a perilously small expeditionary force, he cut matters very fine. As a result of pursuing this policy, which compares with the successful but equally administratively ill-prepared Allied dash for the Meuse in 1944, Caesar has sometimes been accused of having a 'dislike for preparations, due to his eagerness to clinch with his enemy as rapidly as possible'[21] – surely a laudable ambition. But, if the final outcome were victory, which it was, and he carried his soldiers with him, which in the main he did until the divisions of civil war interposed, there can be little justification for such complaint.

Despite the difference in terrain, there are many similarities between the military problems experienced by the Romans in their day, in particular by Caesar in Gaul, and those which confronted Field Marshal Sir William Slim in Burma during the Second World War, when he found himself, in his words, facing three major anxieties – supply, health and morale. There is thus some merit in bearing in mind the difficulties of his XIVth Army, when considering those of earlier Roman commanders.

The strike force, in each case, largely comprised infantry columns, together with lightly equipped local levies. Both were supported, in the main, by the light artillery of their day, frequently pack-borne, and were heavily reliant upon animal transport. Supply was subject to a tenuous system of communication, the bulk being brought forward from the rear, until improved resources, in Slim's case, rendered air supply possible, with some commodities, such as eggs or vegetables, being purchased from local tribes. Rations for as many as seven days were frequently carried by the individual soldier. Scipio's army likewise bore 'several days' rations, under such conditions that they became accustomed to enduring cold and rain, and to the fording of streams'. Quintus Metullus allowed no meat on the hoof, only precooked meat, presumably so as not to slow down the rate of march. In another instance, to quote a much repeated tale, Gaius Marius, in order to reduce the scale of his demand for pack animals, made his soldiers tie their utensils and food up in bundles and carry them on forked poles, over their shoulders.[22] Slim has made it clear to us, in his book *Defeat into Victory*, how he resolved his many problems. Roman generals, with their contemporary biographers, rarely revealed such administrative detail, although it was equally vital to the success of their campaigns.

The health of the soldiers, together with the hospital care and the evacuation of sick and wounded, apart from being morally important, were major concerns of Roman commanders-in-chief, although, obviously, the degree varied according to the personality of the individual. Velleius Paterculus has described how Tiberius looked after the medical welfare of his men during the campaigns of the early first century AD. Throughout the whole period of the German and Pannonian War, he tells us, there was not one of us, of higher or lower rank than ourselves alike, who fell ill without having his welfare and health looked after by Caesar with as much care as though they were the chief occupation of his mind, preoccupied though he was by such heavy responsibilities. There was a carriage ready for those who needed it, his litter was put at the disposal of everyone, and I and others enjoyed the use of this. Now his doctors, now his kitchen, now his bathing equipment, which had been brought for his own exclusive use, relieved the sickness of absolutely everyone.[23]

Others, including such as Hadrian, Trajan and Germanicus were renowned for the attention they gave to the welfare of the sick and wounded, visiting them in hospital and uttering words of encouragement. Indeed, when Germanicus was engaged in fighting, north of the River Rhine, his wife Agrippina, a great-hearted woman according to Tacitus,[24] assumed command of his army's base camp and

Agrippina, wife of Germanicus, 'dispensed clothes to needy soldiers and dressed the wounded'

'dispensed clothes to needy soldiers and dressed the wounded'. The Romans clearly placed more emphasis on medical care than has been chronicled by contemporary military historians, who tend to make more of battles and the dying and the dead, rather than of efforts to succour the wounded. In the days of the Principate, this may have been because of a general awareness that a properly organized military service existed, a consideration well recorded by medical manuals.[25] Frustratingly, they make no mention of its field organization. In the Republican era, it is notable that even Caesar, despite the length of the Gallic War and the many casualties incurred by his army, makes mention neither of his medical arrangements nor of the casualty evacuation system he employed.

Supplies of medical dressing for the wounded were a normal provision in the early first century. All soldiers received some training in first aid, and, within each cohort, the equivalent of a battalion, there was a fully qualified physician with four medical attendants. The importance to morale of providing speedy treatment for the sick and wounded was recognized and, on campaign, tented hospitals were based well forward, deep in the operational area. There is much clear evidence of this, as will be seen in later chapters, in particular in an account by Caesar [26] of an incident in 53 BC, during operations against the Eburones.

Slim considered morale to be a factor of such vital importance that he made it a prime task to tabulate its foundations. Briefly, he considered that, if morale were to be sustained at a proper level, there must be 'some great and noble object' for which a man could strive; that he must have confidence in his leaders and pride in the organization of which he is a member; and that his weapons, working conditions and equipment must be seen to be the best possible. He considered that there was a spiritual as well as a material side to morale, which was a theme also recognized by Montgomery in his *Concise History of Warfare*:[27]

> While operational problems will tend to be the main preoccupation of a general, he must never forget that the raw material of his trade is men and that generalship, basically, is a human problem. . . . The general who looks after his men and cares for their lives, and wins battles with the minimum loss of life, will have their confidence. All soldiers will follow a successful general. A general, therefore, has got to 'get himself over' to his troops. My own technique was to speak to them whenever possible. Sometimes I spoke to large numbers from the bonnet of a jeep, sometimes I spoke to just a few men by the roadside or in a gunpit . . . it is the spoken word above all which counts in the leadership of men.

Both Slim and Montgomery were senior generals, cast in the modern mould and fired in the crucible of the horrendous, mud-filled, blood-stained trenches of Flanders in the First World War. This is where they learnt the principles which were later to guide their actions in North Africa, Normandy and Burma and lead them to the pinnacles of their careers. Their observations on these aspects of leadership therefore demand not only the most careful study but also practice: but it must be said they are not new. Onasander, a Greek military writer of the first century AD, who wrote a treatise on generalship dedicated to Quintus Veranius (who died in AD 59 while in command in Britain), stated his view that a general must be a ready speaker:

> For if a general is drawing up his men before battle, the encouragement of his words makes them despise the danger and covet the honour; and a trumpet-call resounding in the ears does not so effectively awaken the soul to the conflict of the battle as a speech. . . . Should some disaster befall the army, an encouraging speech will give the men's souls new strength; and a not unskilful address by the commander is far more useful in counteracting the despondency of an army in the hour of defeat than the physicians who attend to the wounded.

Onasander also echoed Slim's important message that an army must be motivated by 'some great and noble object':

> The causes of war, I believe, should be marshalled with the greatest care; it should be evident to all that one fights on the side of justice. For then the gods also, kindly disposed, become comrades in arms to the soldiers and men are more likely to take their stand against the foe. For with the knowledge that they are not fighting an aggressive but a defensive war, with consciences free from evil designs, they contribute a courage that is complete; while those who believe an unjust war is displeasing to heaven, because of this very opinion enter a war with fear. . . . For those whose cause is weak when they take up the burden of war, are quickly crushed by it and fail.

Without doubt Caesar practised a similar routine with his generals. It was his habit, dependent upon the operational situation, to address his troops before battle. His accounts of these occasions give the impression that he did this to an assembled gathering but, in view of the large numbers involved and the distances over which they would have been deployed (see Chapter 4), it seems more likely that he rode along the battle line on his favoured war horse, speaking to the cohorts as he came to them, and encouraging the centurions, most of whom he knew personally. He and his generals appreciated the powerful influence of these men and used them as a means of strengthening communication with the troops. In one instance, in the opening stages of the Gallic Wars, and disturbed by signs of disloyalty and a slump in morale among all ranks, Caesar summoned the centurions of every grade to a council, and reprimanded them severely for permitting such a condition to fester. On another occasion, when preparing for operations against the Bellovaci, he summoned his officers and 'communicated to them all the information he had received and told them to encourage the men by passing it on to them'. Likewise his senior general, Labienus,[28] in skirmishing prior to the second crossing of the Rhine, 'summoned the military tribunes and first-grade centurions to explain his plans' to them.

Generally, despite the obvious importance of the subject and with the exception of Onasander, direct considerations of morale, in the modern sense, were rarely uttered by Roman or Greek historians of the day, nor by the ancient military manual writers, such as Vegetius, Frontinus, Arrian and Aenas, who directed much of their work on

The centurions were a hardy, hand-picked body of men of great dependability and courage

the handling of armies in peace and war towards the education of reigning emperors and other senior officers. Truly, Frontinus devoted a chapter to 'restoring morale by firmness',[29] which dealt largely with the lack of leadership by standard-bearers, who were expected to head the advance into battle and sometimes showed reluctance to do so, but this was a narrow approach to a complex subject. Generally, they ignored what Slim called the 'spiritual' factors and stressed, instead, two main components: the essential confidence inspired by training and the need for the strictest discipline, both, of course, highly important. Victory in war does not depend entirely on numbers or mere courage, wrote Vegetius,[30]

only skill and discipline will ensure it. We find that the Romans owed the conquest of the world to no other cause than continual military training, exact observance of discipline in their camps and unwearied cultivation of the other arts of war. Without these, what chance would the Roman armies, with their inconsiderable numbers, have had against the multitude of Gauls? Or with what success, and for the same reasons, could they have opposed the multitude of Germans? The Spaniards surpassed us not only in numbers but in physical strength. We were always inferior to the Africans in wealth and unequal to them in deception and strategem.

He then added the very sage remark that the consequence

Two officer candidates (c. 100 BC) present themselves for recruitment in the Roman army. The clerk on the left is entering the name of one on his ledger. The second candidate appears to be undergoing some form of examination

of engaging an enemy without skill or courage is that part of the army is left on the field of battle, and that those who remain receive such an impression from their defeat that they dare not afterwards look the enemy in the face.

Discipline was ruthlessly, even savagely enforced. Domitius Corbulo who commanded the army in AD 58, during the war with Parthia for the possession of Armenia, was reinforced prior to the campaign by troops from Syria. Tacitus relates that he found the slackness of these troops, demoralized by years of peace-time soldiering, 'a worse trouble than enemy treachery'.[31] Corbulo, a tough and seasoned campaigner, not given to eloquence, determined to provide them with the experience they had missed. He instructed that the whole army would spend the winter under canvas, in weather so severe that ice had to be removed and the ground excavated before tents could be pitched. In the words of the chronicler, frostbite caused the loss of many limbs and sentries were frozen to death. The general was unmoved; thinly dressed and bare-headed, he circulated among the men, encouraging the sick, of whom there must have been many, and praising efficiency. There were, not unreasonably, many deserters: he treated these with unabated severity. Whereas, in other armies, first and second offences were excused, Corbulo executed deserters immediately and without mercy. Tacitus records the remarkable fact that, in the spring, his 'forces were as ready for fighting as for marching'.

On the battlefield, discipline was no less harsh and Frontinus lists numerous examples.[32] When two legions broke in the face of an enemy attack, Fabius Rullus chose men by lot and beheaded them in sight of their comrades. In a similar situation, Aquilius beheaded three men from each of the centuries which had failed to hold their ground. On another occasion, when his defences had been set afire by the enemy, Mark Antony 'decimated the soldiers of two cohorts of those who were on the works, and punished the centurions of each cohort. Besides this, he dismissed the commanding officer in disgrace and ordered the rest of the legion to be put on barley rations.' Human life was cheap and there seems to have been no limit to the numbers of lives taken in punishment. Livy[33] records that, in 279 BC, when Pyrrhus was in southern Italy, the people of Rhegium applied to Rome for protection and were provided with a garrison of a legion which 'criminally took possession' of the city it had been sent to protect. The soldiers plundered the city, killed or expelled the menfolk, and 'took possession of the women and children'. All 4,000 men were condemned to be executed and the Senate declared it a crime to bury any one of them or to indulge in mourning for them.

It is perhaps almost inevitable that, once again, Caesar's approach to discipline, as to so many other matters, should have been different. Perhaps because of his predilection for leading from the front, he understood the chemistry of morale

Trajan addressing his troops before battle

more clearly than most. He judged his men solely by their fighting record. If they failed, he dealt with them all with equal severity. In battle, he kept a careful eye on the ebb and flow of fighting and, if his troops gave ground, he was generally there to rally them. Suetonius relates that he would grasp by the throat any man who fled the ranks and would turn him around to face the enemy.[34] He dealt severely with such behaviour. Discipline in battle was paramount; elsewhere and off-parade, he was inclined to be indulgent. 'My soldiers fight just as well when they are stinking of perfume', he once boasted. As a result, he earned the devotion of his troops to the extent that, at the outbreak of the Civil War, every centurion in every legion volunteered to equip a cavalryman from his savings and the private soldiers unanimously volunteered to serve under him without pay or rations.[35]

Thus, an army must not only be clear in its mind how the complex situations, difficulties and hardships which inevitably arise in war are to be tackled; it must be physically capable of fighting and should be mentally and morally prepared to do so. There is ample evidence to suggest that the Romans were aware of these principles. We may, therefore, also anticipate that running through the entity of the Roman army there was a central doctrine, directing its shape and activities and providing guidelines to be pursued by its general officers. It is thus disappointing, in the various writings provided for us by such as Livy, Tacitus, Josephus, Ammianus Marcellinus and particularly Julius Caesar, including many others, that we are allowed only occasional glimpses of these. Equally, the contributions made to our understanding of Roman military affairs by contemporary Greek and Roman authors of instructional military handbooks have been singularly limited in scope. This is remarkable since much of their work was dedicated to reigning emperors of Rome and, as a consequence, might have been expected to be directed to a more elevated plane than the often simple one it occupies.

Probably the most influential of these writers was Vegetius (fourth century AD), whose work *Epitoma rei militaris*, addressed to Valentinian, was closely studied by European military commanders until as late as the Middle Ages. In presentation and scope, it bears the nearest resemblance, of all its rivals, to a modern training manual. Vegetius had the advantage, of course, of coming on the scene later than most and his work probably benefited from being melded with that of earlier authors. It is said to contain many errors but, nevertheless, this need not necessarily lessen its importance for, without doubt, the work generally points the way to a higher level of military thinking than that of his contemporaries.

Another writer, Frontinus (first century AD), was the author of two valuable military works, namely *The Art of War*, which he followed closely with his handbook, *Stratagems*. This latter volume dealt with military stratagems of Greek and Roman history and won the approval of the emperor Trajan (AD 98–117). Frontinus had been a Roman soldier and, as provincial governor of Britain, had brought about the subjugation of the Silures, a powerful and warlike tribe in south-east Wales. It is thus regrettable, with this professional background – for he could have had much to tell us – that his work, *The Art of War*, should have been lost. It is sometimes suggested that Vegetius may have incorporated much of it in his *Epitoma rei militaris*. Frontinus's *Stratagems* comprises four books, the first concerned with events before the battle, the second with the battle itself and the third with siege warfare. The fourth book is devoted to discipline and there is controversy as to whether or not he compiled it himself. The whole volume contains some nine hundred examples of a variety of 'tactics' employed at different times by commanders in the field and all are selected as being illustrative of various aspects of war.

Some of these exempla are bizarre, one such being an account of an incident during Hannibal's siege of Casilinum, when the Romans scattered nuts on the waters of the Volturnus, as it flowed downstream, so that they might be netted as they arrived in the city and thus beat the food blockade;[36] others are rare nuggets of military practice. Corbulo's dictum, for example, that the pick was the weapon with which to beat the enemy was an expression that might have been voiced by any twentieth-century army commander, even though he was, in this instance, speaking of marching-camp techniques (Chapter 3) and the circumvallation of siegeworks (Chapter 7). Equally, Gaius Caesar's claim that he followed a policy 'of conquering the foe by hunger rather than by steel' has a similarly modern connotation. Vegetius made this latter point even more strongly when he wrote that 'the main and principal point in war is to secure plenty of provisions and to weaken or destroy the enemy by famine'.[37]

The careful attitude of Roman generals to enemy populations, provided they had behaved with reasonable correctness, is also marked by Frontinus. He relates how Augustus, during his war with the Germans, ordered compensation to be paid to local inhabitants for all crops he had included within the fortifications he had erected. In another instance, he tells how a tree laden with fruit, and located inside a marching-camp, 'was found, the day after the withdrawal of the army, with the fruit undisturbed'. And yet again, he writes of a personal experience in Gaul, when the wealthy Lingones people, fearful they were about to be plundered by an approaching Roman army, were so delighted to be left alone that 'they returned to their previous loyalty and handed over to me 70,000 armed men'.[38]

There was a natural wisdom in this exercise in public relations aimed at gaining the friendship of local inhabitants, for Roman armies, compared with the populations of their conquered territories, were small in size, and the numbers of troops required to maintain numerical superiority would have been considerable and wasteful. Thus, as Frontinus indirectly points out, a good relationship with the enemy population was essential. This became progressively more true as the frontiers of conquest advanced and the necessity to return law and order to the occupied homelands became increasingly important. It is possible, here, to make yet another comparison with the war in Burma, where, long before that country's reconquest had been set in hand, Slim had already established a Civil Affairs Organization for this very purpose. As a result, when the relief forces stormed across the Chindwin on their march to Mandalay, teams of civilian officials moved close behind, armed with the essential paraphernalia of administration, and ready immediately to move into the various districts they had already been allocated.

There is evidence to suggest that something of a similar nature took place in the wake of Plautius's invasion of Britain in AD 43. In the preceding years, the tribes situated in the south and south-east of the island had already been cultivated to such good effect by the Romans that, due to the divisions they had created, several princes and princelings had fled to seek their protection. These included Verica, the elderly ruler of the Regni and brother of Tincommius of the Atrebates, and Adminius, the exiled ruler of north-east Kent and younger brother of Togodumnus and Caratacus. There were thus, at the time of the landings, already several British refugees on the continent seeking to return home and re-establish themselves within their community, under Roman patronage. They would have been sufficient in number to form a government in exile. Among them, there was one Cogidubnus, a powerful and ambitious man, an aristocrat, supposedly of British origins and a protegé of Claudius. He had been brought up by the Romans from a very early age and, it may perhaps be judged, had been specifically groomed for this particular moment. Significantly, Tacitus referred to him in his *Annals* as 'an instrument of domination' and 'an example of the long established Roman custom of employing even kings to make others slaves'.[39]

The momentum generated by Plautius's landings at Richborough quickly resulted in the fall of Colchester. This success would then inevitably have been followed by the rapid redeployment of the legions as they consolidated their gains and prepared for further advance. A move of this nature would have required a relaxation of their grip on the territories of the tribes they had defeated in the decisive Medway battle, namely, the Regni, in whose territory lay Chichester, the Atrebates, with their capital at Silchester, and the Belgae, based at Winchester, and would have left the stage open for Cogidubnus to move in and take over their responsibilities. Initially, he would have had the support of the Roman garrison and administrative commander at Richborough, probably Cnaeus Sentius. Ultimately, so it would now appear, he undertook full control, aided by re-organized tribal armies, and was rewarded for his loyalty by a generous grant of 'certain *civitates*'.

Agricola, *Legatus Praetorius* in Britain, AD 78–84, extended these civilian-orientated policies still further during his governorship. According to Tacitus:

> He gave private encouragement and official assistance to the building of temples, public squares and good houses. He praised the energetic and scolded the slack; and competition for honour proved as effective as compulsion. He educated the sons of the chiefs in the liberal arts and expressed a preference for British ability as compared with the trained skills of the Gauls. The result was that instead of loathing the Latin language they became eager to speak it effectively. In the same way, our national dress came into favour and the toga was everywhere to be seen. . . .[40]

Tacitus had no doubt about the purpose of what was happening. The unsuspecting Britons, he tells us, 'spoke of such novelties as civilisation, when in fact they were only a feature of their enslavement'.

In view of the procedure for the selection of Roman generals, with the seeming absence of any recognized military training academies to provide tactical instruction and set uniform standards for achievement, it is not surprising that their military qualities were frequently as varied as their personal characteristics. Many, such as Corbulo, placed an undue emphasis on discipline, which appears to have produced some battlefield success at the cost of disgruntled soldiery. Onasander saw fit to

Trajan's Column at Rome

counsel that 'a general should be alert to the psychology of an army, especially when things are going wrong'; and both Vegetius and Frontinus, in their handbooks, deemed it necessary to include similar chapters of advice, which suggests that such incidents were reasonably commonplace. One of these chapters dealt with 'quelling a mutiny of soldiers' and the other with ways of meeting 'the menace of treason and desertion'.[41] Julius Caesar, perhaps more than any, understood the handling of men, judging them 'by their fighting record, not by their moral or social position' and 'treating them all with equal severity – and equal indulgence'.[42]

Nevertheless, nobody can deny – and the Roman Empire stood as witness – that the system, however strange, produced men of high ability. The key to their achievement lay in the experience which contemporary military writers such as Pliny, Onasander and Frontinus constantly recommended newly promoted senior officers to seek. The renowned Gaius Marius, Cicero relates, 'acquired his military skill on active service and as a commander in wars'. But experience won under such circumstances can be a harsh schoolmaster unless there is a guiding hand. This could only have been provided by the senior centurions, probably the *primus pilus* of the First Cohort, many of whom in due course filled the important appointment of *praefectus castrorum*, camp prefect, and then found the way open to a procuratorship, a civil service staff appointment mainly concerned with financial matters. It is surely unlikely that the military talents of these men would have been then discarded: it is not improbable that they also functioned as staff planners and advisers, providing logistical advice as and when the situation required. A general, advised Onasander,

should either choose a staff to participate in all his councils and share in his decisions, men who will accompany the army especially for this purpose, or summon as members of his council a selected group of the most respected commanders, since it is not safe that the opinions of one single man, on his sole judgement, should be adopted. . . . However, the general must neither be so undecided that he entirely distrusts himself, nor so obstinate as not to think that anyone could have a better idea than his own; for such a man, either because he listens to everyone and never to himself, is sure to meet with frequent misfortune. . . .[43]

CHAPTER TWO

Command and Control

Every operation requires a fixed time for its commencement and a period
and place for its execution. It also demands secrecy, recognised signals,
known persons by whom and through whom it is to be carried out and a
detailed operational plan.

Polybius, *The Rise of the Roman Empire* (*c.* 200–118 BC)[1]

In the twentieth century the demands of two world wars have escalated the
development of battlefield systems of communication with dramatic effect,
emphasizing their importance to victory. Nevertheless, despite the fact that we
now live in a greatly different age, heavily dependent upon advanced technology,
military thinking has changed but little since Onasander wrote his essay on
generalship. The simple principles of war remain undiminished in importance.
Equally, the arrangements made by a senior officer for the command and control
of his forces and the quick transmission of information and instructions to his
troops remain no less vital.

In step with this advance, as it has been extended and refined, the overall speed
of operations has been correspondingly enhanced. Today, orders may be
despatched to the front line and received by formations and units at all levels
almost as soon as they are issued by the commanding general. Thus, a junior non-
commissioned officer in the field, in charge of a half-section (that is to say, in
Roman terms, a half-*contubernium*, or four men), now possesses the ability not
only to remain in constant verbal contact with his superior officer but also to
speak to neighbouring sub-units. Soon, it is probable that every infantryman will
be equipped with a listening device, fitted within his helmet, so that he may be
kept fully aware of the events happening around him.

In the face of such progress, it is easy to forget that, up to the outbreak of the
Second World War, the Royal Corps of Signals still held, within its authorized
establishment, troops of mounted orderlies, or 'gallopers', for the delivery of
messages in the field; and that the British army, even as late as mid-1942, was
engaged in fighting a campaign on the North West Frontier of India equipped,
for communication purposes, largely with signal flags, heliographs and lamps. At
the same time, the remarkably effective practice of passing orders by hand signal
in the field and by bugle-call, both in the field and in barracks, was commonplace.
This latter instance, particularly, was a method well practised, by first the Greek
and then the Roman armies, more than two and a half thousand years ago and

The command group as recreated by the Ermin Street Guard

probably earlier. Asclepiodotus, a Greek philosopher who lived in the first
century BC, a man with no military experience but an avid student of military
theory, has described for us the manner in which they were employed.[2] Although
he has been accused of a dry but admittedly orderly approach to his subject, it is
possible to hear the voice of a twentieth-century soldier echoing the examples
quoted by him (see Appendix 1). It should be the intention of every commander,
he advised,

> to train the army to distinguish sharply the commands given sometimes by the
> voice, sometimes by visible signals and sometimes by the bugle. The most
> distinct commands are those given by the voice, but they may not carry at all
> times because of the clash of arms or heavy gusts of wind; less affected by
> uproar are the commands given by signals; but even these may be interfered
> with now and then by the sun's glare, thick fog and dust, or heavy rain. One
> cannot therefore find signals . . . suitable for every circumstance that arises, but
> now and then new signals must be found to meet the situation; but it is hardly
> likely that all the difficulties appear at the same time, so that a command will be
> indistinguishable both by bugle, voice and signal.

The bugle as a method of communication had its origins with the Etruscans,
the leading bronze founders of the Mediterranean world before the coming of the

Celts. The trumpets they fashioned, and the methods they employed, were later adopted and adapted both by the ancient Greeks and the Romans. Regrettably, neither of these peoples have left us with any clear knowledge of the signal techniques they devised. Nevertheless, there are some slender signposts, firstly erected by Julius Pollux, a Greek scholar of the second century AD and then by Vegetius, an intellectual of the fourth century AD, whose popular textbook, *Epitoma rei militaris*, provides a valuable insight into Roman military practice.[3] The trumpet, or *tuba*, he wrote,

> sounds the charge and the retreat. The cornets are used only to regulate the motions of the colours; the trumpets serve when the soldiers are ordered out to any work without the colours; but in time of action, the trumpets and cornets sound together. . . .
>
> The ordinary guards and outposts are always mounted and relieved by the sound of the trumpet, which also directs the movements of the soldiers on working parties and on field days. The cornets sound whenever the colours are to be struck or planted. These rules must be punctually observed in all exercises and reviews so that the soldiers may be ready to obey them in action without hesitation according to the general's orders either to charge or halt, to pursue the enemy or to retire. For reason will convince us that what is necessary to be performed in the heat of action should constantly be practised in the leisure of peace.

Vegetius, in these paragraphs, lists four calls to be obeyed without hesitation – to charge, halt, pursue the enemy and to retire. From his wording it is possible to infer that a broader range of trumpet calls existed than he has defined. Plainly, whatever range of duties they may have covered, they had two distinct functions: one, to deal with routine administration; the other, broadly operational in nature.

Roman legions, unless on garrison duty, and with the possible exception of campaigns in Britain, rarely operated in isolation but were grouped in army formations. Thus, if their commander had wished urgently to signal an instruction to an individual legion in the field, this could have posed a problem, unless a drill had been contrived for the purpose. Imagine, for example, that an army of four legions was on the march and the commanding general wished to alert a particular legion to prepare for operations. How could its attention have been attracted by trumpet without alerting the whole column? The answer can only be that each legion possessed its own distinctive call, perhaps simply the blast of a musical note to identify its recipient. But what form did this take? Did it convey a musical message by a range of fanfares, each with its own especial meaning and, if so, what would these have been? Or did it simply sound a coded series of notes?

For these purposes, the Romans favoured four particular instruments: the *tuba*, the *bucina*, the *cornu* and the *lituus*. They were rarely employed for musical purposes but, more commonly, on military occasions and for ceremonial duty at civic and religious gatherings.

The *tuba* was an uncomplicated instrument, with a slightly conical shape over its entire length of about 4 feet. It was made entirely of bronze but was sometimes fitted

The cornicen *or horn blower*

with a horn mouthpiece. Without the prefacing call provided by the *cornu*, it was sounded mainly for routine duties. The *bucina* was very similar in size but, in its case, it was crafted with a narrow cylindrical bore, flaring out close to the bell. The *tuba*, readily identifiable by its characteristic sound, served a more important operational purpose. Homer, in his *Iliad*,[4] graphically described how the note of a similar, tuba–like instrument was recognizable across long distances of the battlefield:

> The loud trumpet's brazen mouth from far
> With shrilling clangour sounds th'alarm of war.

The physical effort required to play this instrument was considerable and demanded heavy lip pressure. For this reason, it was the habit of Roman trumpeters to blow with cheeks inflated. Sachs records that in some archaeological representations a chain is shown attached to the bell of the trumpet, presumably to enable the player to brace his lips more firmly against the mouthpiece; and that, in other instances, and for the same reason, the free hand is shown pressed against the back of the head.[5]

The main task of the *bucinator* was to announce the hours of the watch throughout the night and to sound all other necessary calls, as might any modern day orderly bugler, throughout his tour of duty. His role in this connection (almost entirely administrative although he does appear to have been employed to 'blow up' before battle), is emphasized in a 'morning report' papyrus of AD 239, listing the nine men on watch at the standards of *cohors XX Palmyrenorum*, day after day.[6] These included a centurion, three standard-bearers, a priest, a clerk to record the details of the watch, a *bucinator* and two others.

The *cornu* was slightly more melodious. It was made of horn and silver,[7] shaped like the letter G and gently tapered for the greater part of its outstretched length of roughly 11 feet. Horace speaks of its 'threatening rumble', but modern experiments with a facsimile are said to have produced 'a soft and yet voluminous sound'. Its partly circular shape was held together by a crosspiece which rested at an angle on the trumpeter's left shoulder. The tube curved over his head and the bell faced forward, in the manner of a French horn. It is frequently depicted in Roman sculpture, particularly on Trajan's Column. Militarily, the sounding of the *cornu* which, as we have seen, heralded the commander-in-chief's endorsement of whatever instruction was about to be conveyed, carried with it an implicit command for all soldiers to look towards their standards.

The *lituus*, 5 feet long, was a slender bronze tube which curved upwards at its end to form the bell, in the manner of the handle of a walking stick. In contemporary Roman literature its sound was described as *stridor*, a shriek.[8] Its appearance, however, rated better than this, for a surviving example, now in the Vatican Museum, has been described as 'an instrument of great elegance'. The *lituus* is judged to have been used mainly by the cavalry, but its employment is not well documented. It is, however, notable and significant that each of this variety of trumpets possessed its own distinctive sound.

Little is known of the scale of their distribution within a unit but lists of trumpeters of *legio III Augusta* in Lambaesis, which have been unearthed, show totals of 38 *tubicines* and 35 *cornicines*. Using these figures, and on the basis that preserved inscriptions make proportionately less mention of *bucinators* than other trumpeters, it has been estimated that there may have been some 20 of the latter per legion, an average of 2 per cohort.[9]

We have already noted the part played by the commander-in-chief's *cornu*, carried by his personal trumpeter, the *cornicen*. Since the duties of the two instruments were closely linked, he would doubtless also have had a *tuba* in his retinue, with its musician, the *tubicen*. The 'brazen mouth', or carrying range, of this latter instrument would have made it invaluable for passing instructions, over distance, to his formation commanders. Indeed, it is likely that he would have held groupings of both these trumpets, in the manner depicted on the Adamklissi monument,[10] two panels of which show three *cornicines* apiece (thus suggesting a total of six), each man blowing with all his might, so that they might be heard 'from afar', above the noise of battle. Legion commanders, and officers commanding reserve units, may also be expected to have employed both instruments in a similar manner, if not for the use of their formation or unit, then, essentially, for those occasions when they might have been on detached duty or for the purpose of relaying onwards their army commander's messages to neighbouring formations.

The simplest of signals are those which have been preplanned. Polybius[11] demonstrates this when he describes the procedure for breaking camp practised by the Romans:

As soon as the first signal is given, the men strike their tents and assemble their baggage, but no soldier may strike his tent or set it up until this has first been done for the tribunes and the consul. At the second signal they load the baggage on to the pack animals, and at the third the leading maniples must advance and set the whole camp in motion.

The military habits of the eternal soldier change but little, for they are sustained by simple, orderly solutions which withstand the passage of time. The Roman method of dismantling camp provides yet another case in point.

In 1944, towards the end of the Burma campaign, I was adjutant of a Punjabi battalion, encamped with 5th Indian Division on the Brahmaputra plain. We had been engaged on active operations for three years or more and were now being pulled out to return to the North West Frontier. The instructions for breaking camp, contained in the battalion's 'Standing Orders for War', were almost identical to those described by Polybius. We were due to march out at 0800 hours. At 0745 hours, the orderly bugler sounded the battalion call, followed by a G. This was the signal for baggage to be loaded on to the trucks, which stood at the end of every company line. At 0750 hours the bugler sounded a further G: guy ropes were now pulled up, pegs removed and bagged and a sepoy stood inside each tent, holding the centre pole upright. I was standing at the main gate, together with my commanding officer and the Jemadar adjutant, when the third G was sounded. We watched with pride as our camp came down, as if felled by a single stroke. Where, a moment ago, a hundred tents had flowered, providing a home for 600 men, there was now only a flattened field, upon which, once the tents had been bundled and loaded, stood a battalion ready to march. The sergeant of the guard of the British regiment encamped opposite us watched the whole proceeding, expressing his admiration with uninhibited military frankness.

In the face of such similarities, and our lack of knowledge of specific trumpet calls employed by the Roman army, the *Signals by Bugle Horn* practised in 1798 by the newly evolving light infantry are worthy of note,[12] for they demonstrate the potential of this method of communication. The detailed signals described in this handbook were initially listed in a training manual written by a German officer of the 60th Foot, Baron de Rottenburg, and published in that year at the behest of the Commander-in-Chief of the Army, HRH the Duke of York, in 'order that the officers of the army at large may imbibe that degree of military skill and information, which will enable them to discharge their duty to the satisfaction of their superiors, and their own honour, on the most trying occasions'.

There was a total of sixteen calls including several indicating the sighting of a variety of enemy forces, others instructing troops to carry out pre-rehearsed movements, and one for the assembly of officers. In addition to this, the number of calls was greatly increased by the additional sounding of up to three Gs: one, to denote the right of the line; two, to indicate the centre; and three, the left. For example, two Gs sounded before the *Extend*, required the line to extend outwards from the centre; one G sounded before the *Close*, required the line to close on the right flank. Calls were based upon three principles: they should never be resorted to, except when the voice could not reach; they should be as few and uncomplicated as

The range of instructions possible to be passed by bugle (or trumpet) was well revealed by Sir John Moore's Light Troops in 1798

possible; and no movement should ever be executed until the last sound of the bugle horn had died away. Naturally, the weight of firepower with which an army in 1798 was equipped required formations to be deployed over wider areas. Buglers, therefore, addressed smaller units than would have been the custom in Roman times and the instruments themselves had, musically, been greatly developed over the intervening centuries.

The task which confronted Roman trumpeters can best be assessed by considering the circumstances of an actual battle. In 48 BC, Caesar confronted Pompey at Old Pharsalus in northern Greece with 80 cohorts, or 22,000 men under command.[13] Frontinus narrates that

> Gnaeus Pompey [110 cohorts or 45,000 men] drew up three lines of battle, each one ten men deep, stationing on the wings and in the centre legions upon whose prowess he could most safely rely, and filling the spaces between these with raw recruits. On the right flank he placed six hundred horsemen, along the Enipeus river, which with its channel and deposits had made the locality impassable; the rest of the cavalry he stationed on the left, together with the auxiliary troops, that from this quarter he might envelop the troops of Caesar.[14]

Pompey (Cn Pompeius Magnus), a powerful Roman general of the 70s and 60s BC and opponent of Julius Caesar in the Civil War, 50–48 BC

Against these dispositions, Gaius Caesar also drew up a triple line, placing his legions in front and resting his left flank on marshes in order to avoid envelopment. . . .

Caesar stationed the Xth Legion on the right of his line, with the IXth on his extreme left and, since the latter legion had suffered heavy casualties in the Dyrrachium battles, he reinforced it with the VIIIth. He distributed his remaining five legions in between them. He was heavily outnumbered by his enemy and we may thus expect that his lines were no greater than six ranks deep. His cohorts were barely 275 strong, as opposed to an established strength of 500 and Pompey's strength of some 400.

He gave command of his right wing to his general, Publius Sulla, the centre to Domitius Calvinus and the left wing to Mark Antony. Caesar positioned himself, in this instance, behind Sulla on the right wing, from where he could watch the movements of his antagonist, Pompey. He rarely pinned himself to any particular position in the field but generally sought a point of observation from where he might follow the cut and thrust of battle and send help to wherever it might be needed. Caesar quickly observed that enemy cavalry were concentrating together on Pompey's left and, to counter this move, he withdrew six cohorts from his third line and posted them out of sight behind his right wing cavalry and infantry.

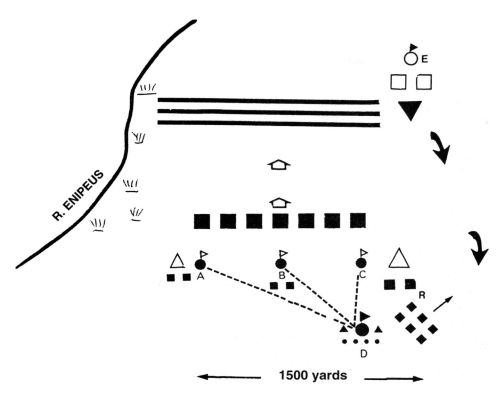

Notes: A. Left wing – commander, Mark Antony. B. Centre – commander, Domitius Calvinus. C. Right wing – commander, Sulla. D. Army commander, Julius Caesar, with supporting arms commanders, staff officers, trumpeters, gallopers etc. E. Enemy commander, Pompey. R. Right wing reserve, with concealed reserve of infantry to counter expected enemy cavalry charge.

Battle at Pharsalus, 48 BC (source: Caesar's War Commentaries, *ed. J. Warrington, III,*

This redeployment would have had no effect upon the width of his front of some 1,200 yards (1,500 yards if allowance is made for flanking troops) which, incidentally, if he had been at full strength, would have reached a distance of 1½ miles.

Caesar's positioning of himself during the Pharsalus battle is of interest. An army commander, according to Vegetius,[15] habitually took up his position in the set-piece battle, with his headquarters, on the right flank of the First Cohort of his right flank legion, and placed himself between this and his flanking cavalry. Customarily, he wore a scarlet cloak (as did Caesar) so that, wherever he might be dragged during the ebb and flow of the fighting, his whereabouts was clearly visible to the troops and, perhaps more importantly, to the junior formation commanders who looked to him for orders. Nearby, and to his rear, he held an élite reserve of horse and light infantry whose role was to manoeuvre into a position from which he could launch a crucial, wide flanking attack upon the right wing of his enemy. The logic behind this tactic was that the enemy held their shields on their left arms, thus exposing their right sides.

We may conjecture that the commander-in-chief's headquarters group was of considerable size. Apart from his personal escort, with standards, trumpeters and mounted orderlies, its numbers would have included mounted staff, specialist officers and liaison officers from his legions, together with others from the cavalry and reserve infantry. Only a selected few of these may be expected to have accompanied him when he chose to move around the battlefield, for too many would have been a hindrance. Mounted trumpeters, for obvious reasons, must have provided important members of this party.

The army's second-in-command was 'posted in the centre of the infantry to encourage and support them' and therefore, we may judge, he had a limited roving commission either side of his post. In case of need, he held a reserve 'of good and well-armed infantry', located clear, and thus to the rear, of the main body, ready to intervene whenever the situation might demand. Almost certainly, he would have been stationed, with his personal staff, standards and trumpeters around him, at the head of these reserves. His role, when the occasion arose, was to thrust through the ranks of infantry lined up in front of him and confront the enemy head-on.

The next officer in seniority, the third-in-command, likewise held in hand a reserve of cavalry and infantry and was posted to command and secure the left flank.

Each of these three men, as has been demonstrated, was stationed several hundred yards apart, in clear, predetermined locations, each with specific individual tasks and his own signal support but looking to his commander for overall guidance. It is thus of interest to compare their positioning, and the signalling requirements we have conjectured, with a similar disposition of commanders found in an action which took place during the crusade of Richard I and which has been recorded for us by Geoffrey de Vinsauf.[16] He relates that, in order to establish battlefield control of the allied force, 'it had been resolved by common consent that the sounding of six trumpets in three different parts of the army should be a signal for a charge', with variations for other phases of combat. In other words, the crusading army was seemingly employing the Roman system of battlefield communication by signals.

The scenario discussed above is not an extravagant view of the communications problem which would have confronted a Roman commander in the field. In the case of an army on the march, with baggage and a straggling animal train, it would no doubt have been greater. When an army was concentrated for defence, as during an incident in Caesar's campaign against the Helvetii in 58 BC, when he was driven to occupy a hill feature, it would have been less. The scale of the deployment of the enemy forces is also a very relevant factor. In 57 BC, when in conflict with the Belgae, Caesar estimated that the light of the enemy watchfires opposing him extended over a front of more than 8 miles. Whatever the eventuality, the solution lay in careful preplanning, an appreciation of the distances over which the trumpet calls were required to carry, the direction of the wind, the shape of the ground and a clear understanding of individual responsibilities within a command structure which, to modern military eyes, was not always as clear-cut as it might have been. Despite this, it proved remarkably effective.

We have already discussed in Chapter 1 the easy movement of top-ranking officers of the Roman army between civil and military posts during the course of their careers. In Republican times, commanders generally were consuls, supreme magistrates at Rome, who were given charge of the army on campaigns. In the imperial army, legion commanders were legates, ex-*praetores* who had in the past held military rank and were now appointed by the emperor to this elevated post. Each legate had a staff of six tribunes who, as we shall see, held a variety of appointments; their tasks were mainly administrative and they exercised no definite military command within the legion. The qualifications of consuls, legates and tribunes have already been well documented[17] and there is no need here to debate them further, except to add that the purpose of this strange mix of civil and military was, of course, designed to keep the hands of the soldiery away from dangerous meddling in political affairs.

Field Marshal Viscount Montgomery, notoriously intolerant of anything he considered to be potentially inefficient, has described the structure of the senior layer of command in the Roman army as 'a military nonsense'.[18] Julius Caesar seemingly held much the same view; he certainly displayed an undisguised mistrust of the worth of his tribunes. When his army, encamped at Besançon in Gaul, became unnerved by gossip about the courage, stature and the high military qualities of the German enemy soon to oppose them, he remarked that the unrest had begun 'with the military tribunes, the prefects of the auxiliary troops, and the men with little experience of war who had followed Caesar from Rome in order to cultivate his friendship'.[19] He appreciated the military risks inherent in nominating short-term politicos to command his legions and obtained authority to select and appoint his own legates. These were tough men with campaigning experience, excited by the opportunity to soldier with him and with no eye cast in the direction of Rome. It must be said, however, that while the early Republican system of appointing senior officers left much to be desired, it was greatly strengthened by the later reforms of Marius (104–102 BC) and Augustus (31 BC to AD 14). Due to these re-organizations, the Roman army achieved a peak of efficiency in the first and second centuries AD.

The mainstay of the Roman army was provided by the centurionate. These men, Polybius has told us,[20] were subjected to a meticulous selection procedure:

> Romans look not so much for the daring or fire-eating type, but rather for men who are natural leaders and possess a stable and imperturbable temperament; not men who will open the battle and launch attacks, but those who will stand their ground even when worsted and hard-pressed and will die in defence of their posts.

In the event, they seem to have found veterans, intelligent, loyal, steadfast and tough, who were prepared to do both.

Customarily, centurions were hand-picked professional soldiers, promoted from the ranks after between fifteen and twenty years' service. There was also a fast-stream system of entry into the centurionate from men of selected property status, but the numbers involved were minimal. In modern terms, it is tempting to regard the centurion as a warrant officer platoon commander, a rank re-introduced in the British army prior to the Second World War but progressively abandoned later, as young commissioned officers became available. Clearly, the position of a centurion was more elevated than this, for a chosen few were regularly promoted to fill the appointment of *praefectus castrorum*, prefect of the camp in peace or war. In other words, the prefect of the camp commanded the modern equivalent of Rear Echelon, with responsibilities for its defence and a full range of other duties, more importantly including the administration of the sick, the maintenance of stores and weaponry, and the re-supply of forward units.[21] Occasionally, he was required to command troops in battle, a fact which underscored both his seniority and his recognized military ability.

The chief centurion of a legion was the *primus pilus*. He had the honour of commanding the first century of the First Cohort and had charge of the legionary eagle; he thus held a vital position in the chain of command, which, we may judge, would have enabled him to influence operational decisions. In the same manner, in Republican times, the senior centurion in each maniple had charge of the maniple standard and the senior centurion of the senior maniple commanded the cohort. In imperial days only vestiges of this arrangement remained for, by that time, centuries were recognized as individual units and the seniority of their respective commanders was strictly enforced.

Vegetius, in *Epitoma rei militaris*, provides a conflict of figures relating to the strength of the First Cohort but agrees, nevertheless, with the generally held opinion that, in the time of the Principate, it was twice the size of the other nine cohorts of a Roman legion. The reason for this differential has been widely debated and it has been speculated that the device may have been initiated to provide a home for the administrative staff and technicians of the legion.[22] This idea is not easily acceptable, for it would have been unusual to find specialists of this nature in the fighting line: their proper place would have been with the baggage train, under the hand of the *praefectus castrorum*.

The First Cohort, as with any other, was essentially a 'teeth' arm unit, which occupied the right of the line of a legion, traditionally the first target of an enemy attack. This flank was vulnerable since the soldiery bore their shields on their left arms, leaving the right side of their bodies unprotected. Additionally, the cohort would have accommodated within its ranks the legionary eagle and other images and, here, in its midst or close by, the legion commander would have stood with his command post and trumpeters. We may judge that the First Cohort of a legion, which had been granted the honour of being posted right of the line of an army in battle, had an even more important role, for it would have been linked to the commander-in-chief and was probably required to provide some of his needs, in the manner of a defence battalion with a modern divisional headquarters.

The standards of the legions had an important role to play in all this, and it is necessary here to consider their significance. To speak of a standard today is, for a layman, to speak of a distinctive banner, the royal standard, the colours of a regiment or, perhaps, the guidon of a cavalry regiment. The *signa* of the auxiliary,

Standard-bearer of XX Legion Valeria depicted by a member of the Ermin Street Guard

legionary and praetorian regiments of the Roman army bore no resemblance to these. The legionary standard, or *aquila*, in Republican times comprised a silver eagle mounted on a thick golden staff, the base of which was provided with a spike so that it might easily be plunged into the ground. Sometimes the staff was fitted with large projecting handles to enable it to be easily plucked out. The bird itself was traditionally shown with a golden thunderbolt gripped in its talons, its wings raised and its head cast forward, displaying its readiness for flight. On occasions, it was featured with its neck strongly bent to the left,[23] as if awaiting authority from the god, Jupiter, with whom it was closely associated, to speed ahead of the marching columns, to seek out and destroy their enemies and secure their advance. Not surprisingly, the Roman soldier regarded his legionary standard with considerable awe.

There is, of course, another reason why the eagle-standards of the legions would have been held in deep regard. In the manner of the colours of a British infantry regiment, surrounded by an aura of past achievements on historic battlefields, they symbolized their legion's valorous past and offered gallant promise for the future. Once consecrated, if they were damaged, however badly, they were not destroyed but repaired again and again, so that their 'life' might be preserved. The loss of an eagle-standard to the enemy, or in any other manner, was a shameful event. In 54 BC, the survivors of Sabinus's army, decimated by the Eburones, fell back to the base camp from which they had come. Lucius Petrosidius, the standard-bearer of the legion, seeing himself beset by a large crowd of Gauls, threw his eagle inside the rampart and died fighting heroically outside the camp. The rest had hard work to withstand the enemy's onslaught till nightfall; in the night, seeing all hope gone, every single man committed suicide. . . .[24]

When a Roman army was on the march, its standards were carried at the head of the legions (there was probably, as we shall see, also a military purpose in this). In camp, they were secured in a shrine of their own, while watch was kept over them as they stood, shaft planted in the ground, beside the altar. It is thus not surprising that they were superstitiously regarded as possessed of unusual powers. Germanicus,[25] so we are told by Tacitus, gained a decisive victory against the Cherusci on the Weser plain in AD 16, after having witnessed the 'splendid omen' of

eight eagles flying towards and into the forest. 'Forward,' he cried, 'follow the birds of Rome, the Roman army's protecting spirits!' The infantry attacked, and the cavalry, which had been sent ahead, charged the enemy's flank and rear.

The enemy host fled in the face of this inspired onslaught, suffering severe defeat, their bodies and weapons being scattered for 10 miles around. A further example of the veneration with which eagle-standards were regarded occurred in AD 41, when Furius Camillus Scribonianus, Governor of Dalmatia, declared civil war against Claudius and called upon the legions under his command to join him. They agreed to do so but the rebellion speedily collapsed when their 'standards resisted all attempts to pull them from the ground'.

In Caesar's day, the eagle was small enough, during one of his adventures, for him to have removed it and concealed it in his girdle.[26] In later years, the legionary standard was made entirely of gold, and the emblem was considerably larger. The staff which it surmounted frequently bore insignia, such as crowns or plates (*phalerae*), which recorded the battle honours and origins of the unit. Vegetius[27] informs us that the ancients

knowing the ranks were easily disordered in the confusion of action, divided the cohorts into centuries and gave each century an ensign inscribed with the number both of the cohort and century so that the men keeping it in sight might be prevented from separating from their comrades in the greatest *melees*. The centurions were distinguished by different crests upon their helmets to be more easily recognised by the soldiers of their respective centuries.

The standard of a maniple, or double-century, also comprised a staff, in this instance topped by an open hand with closed fingers outstretched upwards. It was decorated in vertical array by a wide range of devices representing crowns, half-moons and ships' prows, recording the prowess and origins of the unit in much the same manner as the eagle-standard spoke for the legion. The plates which adorned it, as Vegetius indicates, were probably engraved with the legionary number of the maniple and the number of its parent cohort. Polybius records that each maniple had two standard-bearers, selected from the ranks as being 'two of their bravest and most soldierly men'.[28] For this reason, it has been argued that each century of a maniple possessed its own standard but there is evidence to suggest that each maniple carried two standards: the first, already described, being held rather more for ceremonial duty; the second being a *vexillum*, a small banner of coloured linen, some 50 centimetres square, with a gold fringe on its lower edge and provided with a hem so that it could be affixed to a crossbar on the staff. A standard of this nature would have been much lighter to handle in the fore-front of a battle and, held high aloft, would patently have been easier to identify amid the hurly-burly of the fighting.

Cassius Dio, in his account of the crossing of the Euphrates by Crassus in 58 BC,[29] when the 'portents' were supposedly set against the Romans, provides us with interesting evidence that such an operational banner was carried at army commander level with the purpose, its size would suggest, of enabling his whereabouts to be quickly identified on the battlefield. Additionally, it may have been employed as a substitute for the eagle. Dio relates that the latter

is found in all the enrolled legions, and it is never moved from the winter-quarters unless the whole army takes the field; one man carries it on a long shaft, which ends in a sharp spike so that it may be set firmly in the ground. Now one of these eagles was unwilling to join him in his passage of the Euphrates at that time, but stuck fast in the earth as if rooted there, until many took their places around it and pulled it out by force, so that it accompanied them quite reluctantly. But one of the large flags, that resemble sails, with purple letters upon them to distinguish the army and its commander-in-chief, was overturned . . . in a violent wind. Crassus had the others of equal length cut down so they might be shorter and hence steadier to carry. . . .

From the moment the signalled orders of the commander-in-chief to his legions had been transmitted, the responsibility for passing them onwards and downwards was assumed by sub-unit commanders. They achieved this by what Vegetius describes as the 'motions of the colours', that is to say, the coordinated movement of the various legionary standards, conveying such simple messages as 'halt', 'advance' and 'retreat'. This was common practice until the end of the nineteenth century, when a regiment's colours were still being used as a rallying point, or to indicate a general move into line, or control the rate of an advance or withdrawal. Colonel Galbraith of the 66th Foot, at Maiwand in 1880, attacked by a large army of Afghan tribesmen and holding aloft his regimental colour, rallied 190 officers and men to his side, all of different units, before falling mortally wounded. When he fell, it was snatched once more from the ground and raised high, so that others might see it and be attracted to it. So long as its colours flew, a regiment never died.

There were many Roman similarities to this behaviour, some of which have already been quoted above. In 55 BC, during Caesar's first invasion of Britain, the soldiers of the Xth Legion were hesitant about disembarking until the standard-bearer leapt ashore, holding his eagle aloft, and called upon his comrades to follow him. On another occasion, Caesar, after landing on the North African coast to confront

Labienus at Ruspina in 47 BC, observed that his infantry, which comprised some thirty cohorts arrayed in single line, were becoming disorganized as they ran forward to attack. They were, he noted,[30]

> exposing their flank as they advanced in pursuit of the cavalry too far from the standards, [and] were suffering casualties from the javelins of the nearest Numidians. . . . Accordingly, he had the order passed down the ranks that no soldier should advance more than four feet from the standards.

Thus, Caesar employed his standards to control the pace of his attack. Such a level of achievement could not have been attained unless standard-bearers had been trained to act in concert, to move at the same speed, to maintain a correct distance the one from the other and, we may judge, to dress by the right flank where their senior commander was located, with the First Cohort of his senior legion. By this means, the standards would have held together an unbroken front of advancing troops and would have maintained the line and direction of their assault. More than this, their use in this manner would have eradicated the fault of 'bunching', so often found in attacking infantry.

The fact that Caesar's order not to 'advance more than four feet from the standards' was passed verbally 'down the ranks' is also noteworthy, particularly since, as we have seen, the distances involved were not small. The attention attached by the Romans to this technique has been well underlined by Josephus.[31] In AD 65, as Governor of Galilee, he was preparing an army to resist an anticipated onslaught by Vespasian. He appreciated that his enemy's 'invincible might' was largely due to unhesitating obedience and practice in arms, the result of much training and great attention to detail. He determined to re-organize his army on the same model but recognized that the time available to him did not allow him the opportunity to reach a comparable standard of efficiency. For this reason, he made his mind up to concentrate on essentials. He compiled a brief list of training priorities which included instruction on 'how to pass on signals [and] how to sound the advance and the retreat'. The prominence he attached to the passage of signals is significant for, without doubt, it reflects the consideration given to this skill by the Romans themselves.

The value of pre-planning to simplify the performance of communications and, as a consequence, the tactical or administrative control of a force of whatever size, has already been mentioned. This principle is obliquely underlined in two accounts by Josephus of generals advancing to battle. He provides details of their order of march. In each case, trumpets and standards are grouped together at the head of the column, but his description of the order of march of Vespasian's army, as the latter set out from Ptolemia to invade Galilee, is the most complete and worthy of comment. Behind the vanguard, Josephus narrates,[32]

> rode Vespasian himself with the cream of his horse and foot and a body of spearmen. Next came the legionary cavalry; for each legion has its own troop of 120 horse. These were followed by the mules that carried the battering-rams and other artillery. After them came the generals, the prefects of the cohorts,

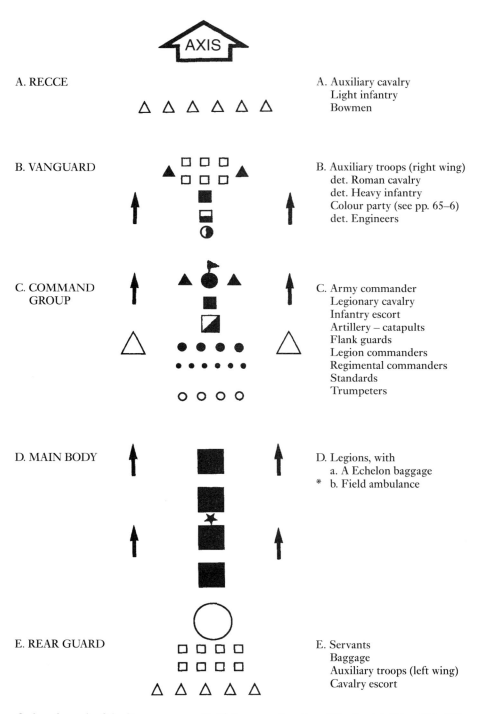

Order of march of the Roman army, AD 70 (sources: Josephus, The Jewish War, *III, 110; Polybius,* Rise of the Roman Empire, *VI, 335; Vegetius,* Epitoma rei Militaris, *III, 78)*

and the tribunes, with a bodyguard of picked troops, and behind them the standards surrounding the eagle which is at the head of every legion. . . . The sacred emblems were followed by the trumpeters, and in their wake came the main body, shoulder to shoulder, six men abreast. . . .

The presence of the generals, so far forward in the column, grouped and separated from their own formations, is significant. It is also notable that they were accompanied by trumpeters and marched in front of the standards 'surrounding the eagle which is at the head of every legion. . .'. It is thus conceivable that they had brought with them their own legionary eagles, standards and trumpeters. If so, this would not have been empty ceremonial, for the Roman army was an essentially practical organization. The eagle, standards and trumpeters formed a vital, functional part within a legion's operational headquarters and it is easy to regard them as inseparable from their commanding general. It is therefore probable that these senior officers, assembled together in attendance on Vespasian, were formed as an 'orders group', a practice strongly rehearsed as a British army battle drill in the opening months of the Second World War and one which remains operative to the present time.

This procedural drill was one by which officers commanding supporting arms, such as cavalry and artillery, and those commanding reserves or sub-units as yet uncommitted to the battle, either moved in proximity to their commander or assembled at a time and place designated by him, so as to be immediately available for a personal briefing whenever the situation might demand. It had many obvious advantages. It saved time and labour in despatching gallopers to call commanders forward; it enabled commanders, when being briefed, to view their allotted task through the eyes of their superior officer; it saved considerable time in assembly; and, if operated correctly, it enabled commanders, before departing to receive their orders, to send a preparatory warning to their units, setting matters in train against their return and yet again increasing the speed of operational movement. Standards and trumpets may be expected to have played their part in such a tactical deployment.

Thus, to sum up: the Roman army exercised tactical command of its troops and control of their movements by the use variously of trumpets, standards and, on occasion, by the passing of verbal messages through the ranks. The trumpets had two distinct functions: one, to deal with routine administration; the other, operational. Operational orders were always prefixed by the sounding of the *cornu*, thus imprinting the commander-in-chief's authority, and were then followed by the relevant call on the *tuba* and a consequent movement of the standards. Due to the mass of soldiers involved, battlefield areas were large and, for this reason or because of adverse winds or unsuitable ground, communications were not always easy to establish. Hence, it is likely that the volume of sound of the instruments was enhanced by employing groupings of them, probably at least six of each type. In the field, they would have operated to a network of relay 'stations', based on the headquarters of formations. It is not unlikely that each formation possessed its own identifiable 'call-sign'. We know little of the manner in which the trumpet calls were employed.

The simplest trumpet calls, as experience has almost invariably shown, are those where action has been pre-arranged. Indeed, it would appear that signals of this category provided the mainstay of the Roman communication system. A medley of calls are mentioned by Caesar in his histories. On one occasion, unable to contain the ardour of his troops in the assault, he claims he signalled them 'Good luck'; frequently he mentions signals to prepare for battle, to charge and to withdraw. Again, in North Africa, with thirty cohorts under command (roughly three legions), he 'gave orders for the line to be extended to its maximum length and for every other cohort to turn about, so that one was facing to the rear of the standards, while the next one faced to their front'.[33] Such complicated manoeuvres were almost certainly pre-rehearsed, if utter confusion were not to result, and, probably more often than otherwise, were executed by command of a trumpet. It is thus more than likely that a much larger range of trumpet calls existed than has been speculated above (see Appendix 2). The simplicity of the technique allowed scope for substantial enlargement.

In the above paragraphs we have discussed the importance attached by the Romans to efficient tactical communications, with the momentum and added control these offered to their military commanders in the field. We have not yet, however, mentioned the parallel strategic communications system which was to be found at government level and which allowed Rome control over the empire's client states and her armed forces in the field. Its prime task was to centralize control of defence and administration and to facilitate the collection of tribute from her dominions. Along an ever expanding network of roads, reaching to her furthest frontiers, this gave an edge to a policy of integrated diplomacy, supported by a broadly unobtrusive, outward-looking, military presence. Indeed, it has been argued that the decay and downfall of the old Republic was, in essence, brought about by the almost total absence of a proper, centrally coordinated communications policy, with the resulting dangerously loose control of ambitious, politically conscious officials.[34] Such communication service as then existed largely rested in private hands and, during the winter months, its operations were generally suspended.

When the emperor Augustus (31 BC to AD 14), ascended the imperial throne, he was quick to see the dangers presented by such laxness and at once set about establishing a national courier service, making progressive improvements to it as he gained experience from its operation. In the early days of his reign, so we are told by Suetonius,[35]

> he kept in close and immediate touch with provincial affairs by relays of runners strung out at short intervals along the highways; later, he organised a chariot service, based on posting-stations, which has proved the more satisfactory arrangement, because postboys can be cross-examined on the situation as well as delivering written messages.

The initial 'pony express' system, and later the chariot postal service, with missives being passed by postboys from hand to hand, enabled exceptionally long distances to be travelled in the course of a day. An example of what was possible is that of Tiberius who, upon the death of his brother, is said by Valerius Maximus[36] to have covered 200 miles in a day and a night; Plutarch[37] relates that Julius Caesar on one occasion travelled 100 miles a day for eight days in succession, driving in a hired *raeda*. Many historians are agreed[38] that a speed of 10 miles an hour, or 160 miles in 24 hours, would have been eminently achievable by the postal service had the requirement not developed for the courier to be cross-questioned on the news he was bringing. The effect of this was twofold. It became impossible for one individual to sustain such a rate of travel, over the distances involved, without adequate periods of rest: and the need emerged for the courier to be an informed staff officer of integrity, capable of making a sensible report. News of the mutiny of the IVth and XXIInd Legions in AD 69, for example, was carried a distance of 108 miles to Vitellius, commander of the Army of the Lower Rhine, by the standard-bearer of the IVth. A delay in transit was considered acceptable providing that someone with reliable background information delivered the despatch.

As the service developed, Augustus insisted that all despatches were signed, sealed and franked with the time of writing. This latter requirement was considered of the utmost importance so that a standard time in transit between places of origin and destinations might be developed for staff-planning purposes. A laurel attached to the despatch symbolized news of victory. A feather affixed to the spear of a messenger, indicating haste, was the insignia of disaster and carried with it an implicit instruction to 'clear the line' and to give the courier all possible assistance.

Procopius, the sixth-century Byzantine historian,[39] provides one of the rare statements defining the speed of travel of an imperial courier. It operated, he writes,

> according to the following system. As a day's journey for an active man they fixed eight 'stages', or sometimes fewer, but for a general rule not less than five. In every stage there were forty horses and a number of grooms in proportion. The couriers appointed for the work, by making use of relays of excellent horses . . . often covered in a single day, by this means, as great a distance as they would otherwise have covered in ten.

This is almost certainly an exaggeration. The yardstick of the postal service, of which Procopius provides us a clue, was that an active man should be able to cover an average distance of 50 miles a day, that is to say a journey of between five to eight 'stages' of roughly 8 miles, the number of stages being dependent upon the difficulty of the terrain to be traversed. Within every stage, forty horses were located at a *mutatio*, a posting-station, together with a proportionate number of grooms, from which relay teams might be selected. At every third stage a *mansio* was located, where night quarters might be obtained by travellers on the road. The distance between each *mansio* was a standard 25 miles.

The Romans made full use of river craft wherever suitably situated (see Chapter 6) but couriers were generally disinclined, if land alternatives presented themselves, to travel by sea because of the inconveniences and uncertainties involved. Ships' captains were reluctant to venture out in bad weather, when passages were uncomfortable and almost always dangerous; moreover, the absence of a compass often compelled sailors to follow the coastline, thus adding to the

Four-wheeled carriage, probably used for fast transportation

journey time. Many journeys, such as the 1,200 miles travelled to Britain by Claudius in AD 43, made use of all the elements of travel.[40] He was aware his presence would be required alongside Plautius when victory appeared imminent. A message would thus have been passed to him by the quickest means, probably a coded signal passed from hand to hand by relays of pre-positioned couriers. It may have taken some ten days to reach Rome. He is known to have set out at once. Sailing from Ostia, Suetonius tells us,

> Claudius was twice nearly wrecked off the Ligurian coast, and again near the Stoechades Islands, but made port safely at Massilia (Marseilles). In consequence he marched north through Gaul until reaching Gesoriacum (Boulogne); crossed the Channel from there; and was back in Rome six months later.

On his northward journey, hampered by the retinue which accompanied him, 'he marched north by road', making full use of pre-positioned animals, supplies, carriages and transport. On his return, it is likely that he made full use of the speed and comfort offered by the River Rhône, on its journey southward to the Mediterranean, through Lyons and the Roman river port at Arles.

A wide variety of vehicles were available for the use of travellers and couriers using the staging service. The most common were the *raeda* and the *cisium*. The former was a four-wheeled covered carriage, capable of fast movement and popular with groups of people with baggage. The *cisium* would have been ideal for courier work. It was a light cart, drawn by two horses, one in the shafts and the other, leading, on a trace, and capable of maintaining a steady 50 miles or so over 10 hours, with one or more changes of horses. The *cisium* was fitted with a single seat, broad enough to accommodate a driver and a passenger. During the late Roman Empire, the *carruca*, a luxurious travelling coach, was developed. It was fitted with a bed on

Two-wheeled cart, similar to some still used in Italy

which the traveller reclined during the day and slept at night. It is difficult to believe, however, that something similar was not available to Claudius on his march to Boulogne, northward from Marseilles.

The postal service envisaged by Augustus was an important instrument of healthy government, with signals and messages flowing strongly in each direction, to and from the extremities of the empire, along a secure and constantly maintained communications network. Procopius, on the other hand, has left us a description of it which emphasizes, rather, an inward, one-way flow, with hints of treachery and insurrection, more in keeping with the collapse of an empire. The system was the brain-child, he wrote,[41] of earlier emperors who

> in order to obtain information as quickly as possible regarding the movements of the enemy in any quarter, sedition or unforeseen accidents in individual cities, and the actions of the governors or other persons in all parts of the Empire, and also that the annual tributes might be sent up without danger or delay, had established a rapid service of public couriers throughout their dominions.

He was, of course, writing in the sixth century, having already glimpsed a vision of an empire in decline.

CHAPTER THREE

Supply Trains and Baggage

The main and principal point in war is to secure plenty of provisions and to weaken or destroy the enemy by famine. An exact calculation must therefore be made before the commencement of the war as to the number of the troops and the expenses thereto, so that the provinces may in plenty of time furnish the forage, corn and all other kinds of provisions demanded of them to be transported.

Vegetius, *Epitoma rei militaris*, iii, 71

However well an army may be motivated and led, whatever heights its standards of discipline and training may reach, unless its resources are adequate, available and properly managed, it will not achieve the victory it seeks. 'Time and opportunity may help to retrieve other misfortunes,' Vegetius reflected, 'but where forage and provisions have not been carefully provided, the evil is without remedy.'[1] This is not to say that there is no room for calculated administrative risk. There is probably no better example of the benefit of such, than the spectacular advance made by the British 21st Army Group which followed the Allied thrust across the Seine in September 1944, after a gruelling three months' fight to break out of its Normandy beachheads.

Montgomery, smarting under the undeserved reputation for slow movement he had gained during the battle of the 'hinge' at Caen, gave orders to his commanders that

> any tendency to be sticky or cautious must be stamped on ruthlessly. . . . The proper tactics now are for strong armoured and mobile columns to bypass enemy centres of resistance and to push boldly ahead, creating alarm and despondency in enemy rear areas.[2]

Second Army's passage northwards, he emphasized, must be 'swift and relentless'. As a result, when British XXXth Corps under Horrocks reached the frontiers of Belgium, their advance (across ancient Gaul, incidentally) had been so rapid and penetrating that they were still dependent on supplies brought by road from Bayeux, 250 miles away. The Americans at Mons and Verdun were even further extended, over a distance of some 400 miles through Paris to Cherbourg, which, at that time, was the only major port in Allied hands. Along the axis of this latter route, the impressive 'Red Ball Express' delivered to United States forces a huge daily total of 7,000 tons of stores. Even so, this load amounted to barely two-thirds of what was needed to keep the American armies advancing. They were grinding to a halt through lack of petrol. Horrocks had similar logistical problems and might have been excused had he acted to seize one of the enemy-occupied Channel ports at Boulogne, Calais, Dunkirk or Zeebrugge. He did not allow himself to be tempted. Instead, he veered away from the coast to thrust directly and successfully for the greater prize of Antwerp, 70 miles distant. The city, with its harbours and strategically important sluice gates, fell into his hands intact.

The advantages were immediate and greatly beneficial. It was an operation of which Julius Caesar would have approved, for so many of his campaigning successes were similarly based on audacious planning and speed of movement, to a degree that experienced military historians have sometimes looked askance at his planning ability. Major General J.F.C. Fuller,[3] whose strong views on the Roman general as a soldier we have already noted, has left us in no doubt of his opinion that, because of this failing, Caesar's armies were inadequately fed. Had it been otherwise, he has argued, why should it have been necessary for Caesar to mention his supply problems to his reader at every turn?

In Gaul, at least, the answer may lie partly in the fact that, when he needed them most to be successful, the crops failed. The harvest in 54 BC was woefully poor on account of drought; two years later, for the same reason, it was little better. On the other hand, although the failure of the harvests may have contributed to Caesar's problems, the reason is more probably to be discovered in the strategy employed by his opponents. The Roman general, frequently deep in hostile territory, was fighting campaigns against a numerous enemy who saw great advantage in disrupting his opponent's supply lines and denying him sources of local food. Caesar records a clear instance of this policy in his *de Bello Gallico*, during the closing stages of the Gallic rebellion of 52 BC. Prior to the last great battle at Alesia, about 30 miles north-west of Dijon, the leader of his enemy, the doughty and gallant Vercingetorix, carefully spelt out to his cavalry officers the lines on which he wished the forthcoming operation to be conducted:[4]

We must strive by every means to prevent the Romans from obtaining forage and supplies. This will be easy, since we are strong in cavalry and the season is in our favour. There is no grass to cut; so the enemy will be forced to send out parties to get hay from the barns, and our cavalry can go out every day and see that not a single one of them returns alive. What is more, when our lives are at stake, we must be prepared to sacrifice our private possessions. Along the enemy's line of march we must burn all the villages and farms within the radius that their foragers can cover. We ourselves have plenty of supplies, because we can rely upon the resources of the people in whose territory the campaign is conducted; but the Romans will either succumb to starvation or have to expose themselves to serious risk by going far from their camp in search of food.

These, of course, were not the precise words used by Vercingetorix: in the custom of many ancient writers of the day, they were probably put into his mouth by Caesar and represented the military dilemma by which he was then confronted. It was Caesar's problem that his enemies, operating from the strength of interior lines, exploited a vigorous scorched-earth policy, making maximum use of their superior numbers of cavalry to deny him supplies, and constantly threatened his baggage trains. The loss of the baggage, with its valuable contents, if it had occurred, would have dealt him an unpleasant blow. It was his task to ensure that this did not happen.

However much care may be devoted to its defence, an army on the march is a vulnerable entity. Today, with the protection of fighter aircraft overhead, with mechanized troop carriers, speedy bulk-carrying road transport and all the advantages of air supply, important reserves of *materiel* and men may be retained in safe areas, distant from the immediate dangers of the battlefield, until required to be called forward.

Such an arrangement was rarely feasible during Caesar's Gallic campaigning, when control of local supplies was his ambition but often beyond his grasp. Of necessity, he was compelled to seek help from uncertain allies to feed his troops in the field and, simultaneously, build up reserves of grain to satisfy the demands of the coming winter season. He retained these reserves, together with any surplus baggage, within a nearby firm base, in towns such as Amiens, Nevers or Sens-sur-Saône, mostly riverine settlements it should be noted and generally located within Aeduan territory. He would then provide it with the protection of one or two of his weaker legions, in essence thus creating a rear headquarters. Now and again, according to the situation, he gave its command to a dependable general, such as Labienus, who regularly acted as his second-in-command. On other occasions, he employed a senior *quaestor* for this duty, an appointment broadly comparable in rank to that of quartermaster-general. Whoever he may have selected, it was invariably a man in whom he had wholehearted trust.

Two noteworthy examples of this tactic provide us with an idea of the wide variety of 'baggage' involved and also, despite the protection awarded to it, of its vulnerability. The first incident,[5] recorded in 54 BC, occurred when the Nervii launched a surprise and devastating attack upon Cicero's winter camp. Caesar determined, at once, to march to the assistance of his brother officer who, within a few years, was to be embroiled in a civil war, fighting alongside Pompey, against his commander-in-chief. Before departing from his Amiens headquarters, Caesar called forward Marcus Crassus, encamped nearby with his legion, and instructed him to take command, in his absence, of everything he was leaving behind, 'the heavy baggage, the hostages furnished by various tribes, the state papers and the whole of the grain that had been collected to last through the winter'.

The second example took place two years later, when the Aedui, until recently Rome's ally, raised the flag of rebellion and declared their allegiance to Vercingetorix. This happened at a moment when Caesar had already disposed of his heavy stores, with other encumbrances, and had placed them under military guard at Nevers, an Aeduan town situated in a strategic position on the River

Loire. The 'heavy baggage' on this occasion is listed as including all the hostages he held from the various Gallic states, as well as his stores of grain, public funds and archives, a large part of his personal baggage, with that of his troops, and a number of horses he had bought in Italy and Spain for use in the war. The Aedui took uncompromising advantage of Caesar's absence. They massacred the garrison, together with some Roman merchants, probably traders marching with the legions; they shared out the money and the horses; they carried away as much grain as they could transport and threw the remainder into the river; and they handed over the hostages to magistrates at Bibracte, an Aeduan settlement on nearby Mount Beuvray, 12 miles west of Autun.

Almost certainly, these lists of 'baggage' are substantially understated, for it would have been normal for many other items of men and *materiel*, not listed by Caesar, to have been found at a headquarters of this importance. These would have included ancillary troops, with numerous clerks, technicians and specialists, together with other items such as reserve tentage, weaponry, clothing and cavalry equipment: a field hospital, with medical staff and accompanying sick; probably a training ground, or *gyrus* for cavalry horses, with associated veterinary staff (*veterinarii*); engineering stores and bridging equipment; and artificers' workshops,

Trajan's 'artificers at work during battle'

in addition to those, or detachments of them, which may have been located forward, travelling with the main body of the field army.

Reserves of tentage and other associated stores not required on the line of march would have been substantial. Roman officers were uninhibited about the exercise of privilege; the tent of a centurion, in area 20 feet square, was twice the size of the *papilio* designed to house an infantry section of eight men. The tents of tribunes and others of comparative rank were taller structures, carried on box-like frames of poles and slats and paved with cut turf, and constructed to accommodate a dining table and couches. The marquee of a commanding general, occupying ground space 200 feet square, was likened by Josephus to a temple. Julius Caesar, always a man of character, impressed his guests by carrying around a mosaic floor in portable sections,[6] which he presumably used to embellish his tent in winter quarters.

Thus, many uncertainties surround the size and composition of a Roman baggage train. A prime example is the fluctuating ration situation, created by the success or otherwise of foraging expeditions, with its effect upon the transport requirement. How should one compensate for this? How many servants would have travelled with the train and how were they employed? How many sick? How many ambulance wagons? How many hostages, if any, with their attendants? How many tradesmen moved with the column, always ready to exchange money or barter for loot? How much loot, accumulated by officers and men, and what scale of transport should be allowed for it? And many other questions similar to these, all of which would affect our calculations. Now and again, in *de Bello Gallico* and

Oxen, as well as horses, were used for hauling transport carts carrying water and wine

other writings, there are lines which contain the hint of an answer to some of these points, but clear evidence remains elusive.

One such passage is to be discovered in Caesar's account of the fighting in 58 BC, a year which found him campaigning in eastern Belgium, north of Liège. He had separated his army into three divisions of three legions each and, commanding one of these himself, he despatched them on a widespread foray, probably a reconnaissance in force, combined with a search for grain, forage and loot. He gave orders for all to re-assemble within seven days at the old Roman camp at Atuatuca, the fortifications of which were still intact. He had selected it for this very reason and, here, he now deposited his heavy baggage. He provided the XIVth Legion to watch over it and placed Cicero in overall command, warning him not to risk leaving camp due to his limited strength. It proved an uneventful week, and the soldiery left behind were disgruntled at being confined within the fortification for apparently so little purpose. Then, when the return of the legions was expected within a matter of hours, Cicero released them. He was convinced, so we are told, that there was no good reason why they should remain cooped up any longer;[7] there was, of course, the added incentive that he was nearly out of food and forage:

> Accordingly he sent five cohorts to get corn from the nearest fields, which were separated from the camp by only a single hill. A number of sick men had been left behind by the legions: from these there went with the cohorts, as a separate detachment, some three hundred men who had recovered from illness during the week, and permission to accompany them was granted also to a large number of servants, who took out a great many of the animals that were being kept in the camp.

Some of the detailed facts we are seeking are concealed in these words. The 'number of sick', of whom by the end of the week some three hundred had recovered, makes it likely that, at the time of Caesar's departure, there was an overall total of some four to five hundred casualties in the camp, a figure which would suggest the presence of a field clearing station, with carts, wagons and animals for medical stores and ambulances. But of even greater interest is the mention of 'a large number of servants' who, 'with a great many animals', left the fortification to accompany the cohorts to nearby fields, doubtless seeking fodder. The number of servants, their origins and the manner of their employment, is a subject to which we return below, for it contains a considerable and largely unappreciated manpower commitment.

The vulnerability of a marching column increases in direct proportion to the size of its baggage train. If the latter is needlessly unwieldy, it makes extravagant demands upon defence resources and detracts from the strike power of the fighting troops. It is an intriguing comment that Caesar, although he frequently mentions the wheeled transport operated by the enemy, rarely writes of using any himself. It may be that, in his yearning for surprise, speed and mobility, he depended upon the pack mule, at least for his first line transport. If so, he would have held the same view as Philip of Macedonia, who forbade wagons to be used by his army.[8] Philip, as well as Alexander the Great, saw them as a brake upon his movement and a tempting dumping place for unauthorized baggage, thus adding to his administrative burden. Indeed, Alexander, in his great march across Iran and Afghanistan to India, ordered all carts and their surplus contents to be burnt, beginning with his own. However, this is not to say that Caesar would have behaved as radically as this, for his light-weight general duty carts were employed in many valuable ways and were, moreover, capable of moving at 4 m.p.h.,[9] a speed which would have more than competed with that of the average infantryman.[10]

The logistical needs of Caesar's army in Gaul were a direct reflection of its size and composition. It is therefore necessary, before we go further, to determine how large it might have been.

In 58 BC, in the weeks prior to his campaign against the Helvetii, Caesar hastily put together a force of six legions, of which three had been summoned from winter quarters at Aquilea, a town in Cisalpine Gaul, at the head of the Adriatic; two had been newly raised in Italy and one was already occupying a defensive position, under the ubiquitous Labienus, between Geneva and Jura. Within a short time he increased this number to eight and his infantry corps remained virtually unchanged at this level throughout his campaign, except for its closing months when he called together fourteen legions to give him the decisive victory which, by then, he so desperately sought. He regularly deployed two of his less experienced legions to safeguard his rear headquarters, with its supply base.[11]

Caesar's cavalry wing was rarely more than four to five thousand strong and

A Roman auxiliary cavalryman: note the absence of stirrup irons

was mainly raised from Provence, from his allies the Aedui and other minor tribes associated with them. Occasionally, he acquired the additional help of a few hundred high-quality German cavalrymen. In the manner of the Pathans on the North-West Frontier of India in more recent years in their wars with the British, they fought for him with as much ferocity as they occasionally fought against him.

Fuller,[12] who queried the adequacy of Caesar's supply arrangements, has also questioned why Caesar, throughout a campaign which lasted several years, felt compelled to requisition between four and five thousand Gallic horsemen from chieftains whose good faith he held under grave suspicion. The answer, as previous Roman commanders would have discovered, for since the days of Marius (d. 86 BC) cavalry had ceased to be recruited from Roman citizens, is almost certainly logistical. By raising cavalry support in this manner, he was saved the tasks of recruiting, training and providing equipment; the tribesmen provided their own horses and doubtless found their own remounts; and, during the winter months, being generally unrequired, they returned to their homes and were removed from the ration list. Nevertheless, as events proved on at least one occasion, they were liable to recall.

There was also a hidden bonus. Horses, particularly cavalry animals, were ridden hard during the campaigning season and, as is normal in war, where days of plenty intermingle with days of famine, we may expect they received scant rations and were in poor condition on their return home. By this means of raising

Cavalry scouts depicted by the Ermin Street Guard

cavalry, with its 'yeoman' flavour of recruitment, the responsibility of coming back next year with an animal of renewed fitness now rested with the owner of the horse, and the Roman *quaestor*, upon whom it might otherwise have devolved, was relieved of an onerous and expensive administrative task. The savings in men, money and *materiel* must have been considerable and it would not be surprising to learn that Caesar filled many of his other transport needs in a similar manner.

For maximum efficiency, Caesar's transport would have been divided into three operational echelons:

- A Echelon, containing first line transport such as wagons (if any), carts and pack animals supporting the army in the field.
- B Echelon, or second line transport, such as wagons, carts and pack animals (both of the latter as required) operating between rear headquarters and the army in the field, ferrying supplies and stores to forward areas and returning with wounded, salvaged weaponry etc., in otherwise empty wagons.
- C Echelon, or third line transport, mainly grain wagons, plying between rear headquarters and/or the field army. River transport may also have been found in this category (and occasionally with B Echelon).

First line transport requirements of a legion are imprecisely known but we can arrive, with a broad degree of accuracy, at a total of some 1,250 horses/mules per legion for non-comestible items. This calculation is based on a scale of one pack animal per section of eight men (*contubernium*),[13] carrying its tent, mill, kettle, with other items of kit and tools; one two-wheeled cart per century, drawn by two horses or mules, for the artillery piece (*carroballista*); another two-wheeled cart, also drawn by two animals, for each century, perhaps to carry the centurion's tent and luggage, but possibly to take unwanted equipment from the soldiery when fighting was about to commence. To these totals should be added a detachment of 120 Roman cavalry (the normal legionary complement) and some 250 animals to cover the demands of staff officers, the ambulance and other supporting services. For an army of six legions, this figure would multiply to 7,500 horses or mules.

In order to calculate the total of first line pack animals required for the carriage of grain, without taking account of the requirement for animal feed, we need to be told how much would have been carried on the man and how much would have moved with column reserve. Livy, in an account of the war against the Aequians,[14] describes how soldiers brought with them five days' 'bread ration'. Let us, therefore, take, for purpose of discussion, an arbitrary allocation of five days' rations carried on the man, with a further ten days' reserve carried by pack transport. This would have called for 15 lb of grain to be carried by each man, with an additional and, it must be said, an improbable 5,625 pack animals to carry the column reserve. A more feasible solution would have been for each man to carry ten days' rations (30 lb), and for a further two days' reserve to be carried by pack transport. If these latter rations were the first to be consumed, an important reserve of 1,250 pack animals (the equivalent of 112 wagons)[15] would thus have been created for the transport of 'windfall' grain stocks foraged from the surrounding countryside.

It will be evident that these figures are, as yet, incomplete; nevertheless, it is already apparent that baggage animals in this quantity, moving through potentially hostile territory and with precious cargoes, would have been very likely to scatter in an emergency unless they were tightly controlled (see Appendix 2, note (e)). The majority of their loads would have been specialized, frequently needing quick identification, and for them to be even temporarily lost amid several thousand animals would have been administratively intolerable. Moreover, if the army commander was looking for a speedy and disciplined advance, then they must have been led by hand. The task of a muleteer has been described by the joint authors of a splendidly humorous book, *Jungle, Jungle, Little Chindit*, published in 1944. Any soldier who served in the 1942–45 Burma Campaign will readily identify with its sentiments and vouch for its truth:[16]

> The mule man, mule-leader or muleteer looks after one mule, its saddle and harness. He must feed it, water it, wash it, groom it, keep its outfit clean, see that its load, when on, is properly tied and balanced, prevent it from becoming galled, broken-winded or lost, haul it up when it goes rolling down the mountain side, run after it when it runs away and take care that in harbour it stands in a space carefully cleared of rustling leaves and does not stamp on the face of nearby sleepers. Apart from these little jobs, and a certain amount of marching and fighting and attending to his own personal needs, his time is his own. . . . He will find that his mule responds kindly to attention and that mules, especially when tired, are not necessarily stubborn and do not necessarily kick (at least not very hard).

But whose was the responsibility for this task? Was this the role of the servants? It will be recollected that Caesar, recounting the ambush set up by the Eburones at Atuatuca, remarked that servants authorized by Cicero to leave camp took 'a great many' animals with them, presumably to forage for fodder and grain. Vegetius, in his *Epitoma rei militaris*,[17] likewise associates servants with the task of handling 'baggage' animals. The ancients, he writes,

> were very careful that the servants or followers of the army, if wounded or frightened by the noise of action, might not disorder the troops. . . . They ranged the baggage, therefore, in the same manner as the regular troops under particular ensigns. They selected from among the servants the most fit and experienced and gave them command of a number of other servants and boys, not exceeding two hundred, and their ensigns directed them where to assemble the baggage and the troops. . . .

Julius Caesar, on the other hand, has described an incident which occurred during his conquest of the Belgae in 57 BC,[18] when he was setting up a marching-camp under difficult conditions. The servants, he tells us, 'had gone out to plunder', deeming that all was safe to do so, when they were alerted to danger by 'shouting and din [which] arose from the drivers coming up with the baggage'.

From these words, and the remarks of Vegetius above, we may judge that servants and drivers alike were both involved with the task of animal handling, the former because of their duties as servants to the military, the latter in a more general manner, perhaps as drovers of groupings of animals. As far as the servants are concerned, we can look for an example of their employment to Philip of Macedonia, justly famed in ancient times for his strict logistical discipline. He not only forbade the use of wagons in the army, but restricted numbers of servants to a scale of one for every ten infantrymen and one for each cavalryman, in the latter case for the purpose of carrying his hand mill and other equipment. Presumably this freed the horseman for the important task of looking after his animal and for other duties. If we take this as a yardstick, therefore, it would seem reasonable to assume that each *contubernium* (eight men) was provided with a servant to look after the section mule, making a total of 600 servants for this purpose within a legion, regardless of the needs of other ancillary troops,[19] making an estimated total of some thousand servants per legion on the basis of one servant per animal. The drivers, as general duty handlers with no other specific role, in a modern army would probably have been allocated two pack animals per man,[20] but in this instance would probably have been expected to cope with many more (but see below and Chapter 4, pp. 75–7).

The final requirement in assessing the size of a baggage train is the difficult matter of the horse feed. In peaceful times, a horse or mule may be expected to consume some 20 to 24 lb of dry fodder a day, but this would have been an unlikely scale in the field where hard feed, under combat conditions, would have been difficult to find.[21] At the very least, the animal would have needed five hours' or more good grazing, depending upon its temperament, always assuming the grass was sufficiently lush. Moreover, with the large numbers involved, the area of grazing would have needed to be extensive. Without doubt, feeding their animals on operations must have presented Roman planners with a considerable problem and their answer to it, without adding considerably to an already high transport commitment, is not easy to identify. An inkling of a solution is revealed by Caesar when he tells of the administrative difficulties he overcame in his pursuit of the Helvetii in 58 BC.[22] Due to the cold weather, grain in the fields had not yet ripened and there was an insufficient supply of hay. The Aedui, who had undertaken to re-supply him, had failed to do so and, an added complication, his enemy had swung away from the river rendering 'the grain that he had brought up the Saône of little use' to him. It thus appears that Caesar's second and third line transport, suggested above, justifiably included river craft.

This is a very significant revelation for, if he were making use of river transport (and, from this evidence, he was) then many of his logistical problems would have been resolved, including, importantly, the delivery of grain and hay for those of his animals which, by nature of their duty, were pinned to the column on the march. Linking the column with such a river supply base could have provided a task for the 1,250 reserve 'ration' mules we visualized above. The cavalry, on the other hand, widely deployed in a protective screen around the army on the march, and penetrating deeper into the surrounding countryside than the transport animals would have found possible (they could have covered 40 miles in a day) would have had ample opportunity to forage for themselves.

Relief from a funeral monument found at Neumagen, in Germany, showing barrels of wine being transported in a rowing barge. The steersman is clapping his hands, beating time for the oarsmen

But what were the sources of Caesar's grain supply and how did it reach him? To find the answers to these questions we must return to the commencement of his campaign.

Rome, for some years before Caesar's invasion of Gaul, had established friendly relations with the tribes of Transalpine Gaul, particularly the Aedui, an important tribe, but one whose influence was beginning to wane. In 61 BC the Aedui were defeated in battle by their neighbours, the Sequani, and, in the ensuing peace negotiations, were persuaded to cede to their enemy the right, which until now they had enjoyed unquestioningly, of collecting tolls on the River Saône. They appealed to Rome, as a result of whose intervention their privileges were quickly restored and the friendship which already existed between the two peoples was thus strengthened. In 58 BC Caesar was already looking ambitiously in the direction of Gaul, when the great western emigration of the Helvetii in that year provided him with the opportunity he was seeking, for the Helvetii, having passed through the Sequani homeland, now invaded Aeduan territory. The Aedui, 'unable to protect themselves or their property . . . sent to ask Caesar for help, pleading that they had always been loyal to Rome and that it was not right to allow their land to be ravaged almost under the eyes of his army'.[23]

Caesar quickly disposed of the matter of the Helvetii and next marched to the aid of the Sequani, whose territory was suffering under occupation by Ariovistus, a German warlord with clear ambitions to extend his influence south of the Rhine. He was persuaded by Caesar to withdraw and the latter then turned to cast his eyes elsewhere. As their Roman ally's appetite for conquest increased and his demands for supplies grew heavier, the Aedui began to question whether it was not better to be under the heel of a fellow Gaul than that of a Roman conqueror. Caesar complained that the Aedui kept putting him off from day to day, saying that grain was being collected, was in transit, was on the point of arriving, and so forth. When he saw there was going to be no end to this procrastination, and the day on which the soldiers' rations were due was approaching, he summoned the numerous chiefs who were in the camp . . . [and] reprimanded them severely for failing to help him at such a critical moment, when the enemy was at hand and it was either to buy corn or to get it from the fields.[24]

Despite their evident doubts, the Aedui and the Sequani, particularly the former, appear to have remained Caesar's principal suppliers, although, during the ebb and flow of the war, there were others further afield, such as the Veneti, who were occasionally called upon to provide this levy. The Aedui and Sequani appear not only to have transported supplies to Roman rear bases, but also to have delivered direct to the army in the field. This becomes evident during Caesar's narrative of his encounter on the Alsace plain with Ariovistus. The two armies were encamped, confronting each other, when the German general emerged from his defences to march his army around Caesar's assembled legions, and take up position 2 miles further on, 'with the object of intercepting grain and other supplies that were being sent up from the Sequani and the Aedui'. Again, in 52 BC, during the opening weeks of the rebellion led by Vercingetorix,

Caesar advised the Aedui to forget their disputes and quarrels and allow nothing to distract them from the war they had on hand. They might look forward to receiving from him the rewards they deserved when the conquest of Gaul was completed. In the meantime, they must send him without delay their cavalry and ten thousand infantry to be distributed at various places for the protection of his convoys.[25]

We can only conjecture the extent to which the Aedui used river craft to supply Caesar's legions but the location of their territory, adjoining the Auvergne region and largely comprising present-day Burgundy, would have rendered them well sited to exploit the far-reaching waterborne transport system this offered to them. The *departements*, which today provide the boundaries of this fertile wine-growing district, read like a gazetteer of famous French rivers, with their tributaries. To the north lies Seine-et-Marne, Aube and Haute-Marne; to the east Haute-Saône, Jura and Ain; to the south Rhône, Loire and Allier; and to the west Cher and Loiret. Thus the Aeduan homeland was not only prosperous and fertile, it was also one of considerable strategic importance, because of its location as well as for the great many rivers which fanned out from it, flowing westward to the Channel and southward to the Mediterranean Sea. Their employment of the waters of the Saône has already been mentioned. It would have been strange indeed if this important facility had passed unnoticed by the astute Julius Caesar in his journey of conquest.[26]

The vulnerable and unwieldy nature of a baggage train is self-evident. Nevertheless, it must also be admitted that, being unaware of a clear method of

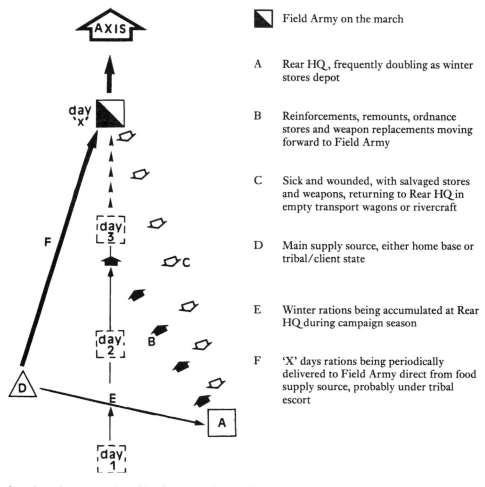

A   Rear HQ, frequently doubling as winter stores depot

B   Reinforcements, remounts, ordnance stores and weapon replacements moving forward to Field Army

C   Sick and wounded, with salvaged stores and weapons, returning to Rear HQ in empty transport wagons or rivercraft

D   Main supply source, either home base or tribal/client state

E   Winter rations being accumulated at Rear HQ during campaign season

F   'X' days rations being periodically delivered to Field Army direct from food supply source, probably under tribal escort

Supply technique employed by Caesar in Gaul, 58–51 BC

handling it, we tend rather to look at the great number of animals and stores which comprised its content and to ignore the discipline of organization which the Roman military command must have imposed upon it. Vegetius provides us with a hint of its shape when he wrote his line that the 'ancients ranged the baggage in the same manner as the regular troops under particular ensigns'. The latter, each in charge of a specific grouping, would have looked to the *quaestor* for their instructions. He, in his turn, would have received his orders from the commanding general. From this we may judge that, when transport was massed on the march or for active operations, legions and legionary sub-unit groupings would have been kept intact, even down to cohort level, if not below. By this means, the transport and baggage of any sub-unit could readily have been detached for duty with it should this, at any time, have proved necessary.

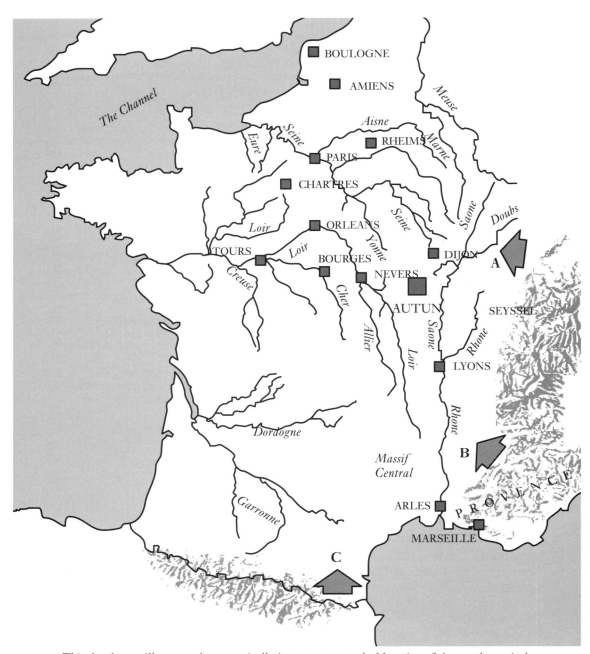

This sketch map illustrates the strategically important watershed location of Autun, the capital city of the Aedui, Caesar's allies and main food supply source. The arrows indicate the traditional invasion routes into Gaul. Caesar's legions entered Gaul broadly along the axis of arrow A

TABLE 1
BAGGAGE ON THE MARCH

			Pack Animals		Distance (a)	
			Legion	Army (b)	Legion	Army
REAR GUARD:						
!.	Carts					
	– artillery 60					
	– centuries 60					
	– sundry 30	Total 150	300 (c)	600	750	1,500
2.	Ration reserve		125 (d)	1,250	625	3,125
3.	General Duty		950	7,500	2,375	19,000 (e)
	Total		1,375	9,350	2,875	23,625 (f)
MAIN BODY (b):						
4.	Carts					
	– artillery 60					
	– centuries 60					
	– sundry 30	Total 150	300 (c)	1,800	750	4,500
			1,675	11,150	3,625	28,125

NOTES: (a) Distance shown in yards: estimated 15 feet per animal, 30 feet per cart: animals and carts two abreast. (b) Based on an army of six legions with main body, two with rear guard. (c) Two animals per cart. (d) Estimated allocation of ration reserve. (e) GD transport for eight legions. (f) Rear Guard transport approximately 13.5 miles: Main Body transport 2.5 miles: servants/animal handlers included in space allowed for animals: overall transport 16 miles.

TABLE 2
FIELD ARMY'S GRAIN REQUIREMENT FOR ONE DAY (a)

	Numbers	Rations	Animals	Grain (b)
Personnel	37,500	3 lb		50
Cavalry				
– Gallic (c)		3.5 lb	4,000	6
– Roman (d)		3.5 lb	720	1
Reserve animals		3.5 lb	1,250 (e)	2
Baggage animals		3.5 lb	7,500 (e)	12
Servants	6,250	3 lb		8
Total	43,750	20 lb	13,470	79 tons

NOTES: (a) Cavalry wing 4,000, legionary troops 30,000, ancillary troops 3,500, Total 37,500. (b) To the nearest ton in each case. (c) animals only: see note (a). (d) But cavalry horses could well have been self-sufficient: see text p. 50. (e) Total baggage animals = 8,750.

Table 1, which is based on the baggage needs of an army of six legions, with ancillary and supporting troops, and consolidates the figures we have already discussed in the above paragraphs, reveals a conservative requirement of 8,750 transport animals. In these calculations some major assumptions have been made, namely, that the cavalry would have provided their own grain and forage; and that transport animals would either have found adequate grazing and foraged grain, or that feed would have been brought forward to them from the army's rear (perhaps riverine) supply base. Table 1 thus provides a minimum situation and reveals that the baggage element of an army of this size, with transport animals marching in pairs and carts moving singly, would have extended over a distance of some 12½ miles, a figure which compares reasonably with other notable instances. In the Boer War, for example, General Oliver's army of 6,000 men, together with its baggage train, is said to have extended on the march over a distance of 24 miles.[27]

It is thus not surprising, for this reason and because of the military value of the supplies they carried, that baggage trains provided such attractive targets for the enemy and were so keenly defended. A typical incident occurred in 52 BC, when Caesar was marching through the territory of the Lingones to the support of his provincial troops. Three divisions of cavalry under the Gallic leader Vercingetorix, who had been watching his progress carefully, descended upon his column. Of these, two threatened his flanks, while the third took up a position across his path to deny his vanguard further advance. Caesar split his cavalry up in a similar manner and

> ordered them to advance against the enemy. Simultaneous engagements took place all along the column, which halted and formed a hollow square with the baggage inside. If Caesar saw the cavalry in difficulties, he moved up some of the infantry and formed line of battle, which hindered the enemy's pursuit and encouraged the cavalry by assurance of support. . . .[28]

When we recognize the size of the baggage train, together with the distances involved, we can only admire the standards of training and the obvious military skills which permitted such a coordinated and controlled response.

CHAPTER FOUR

Marching-camp Techniques

The importance of [fortifying camps] appears not only from the danger to which troops are perpetually exposed who encamp without such precautions, but also from the distressful situation of an army which, after receiving a check in the field, finds itself without a retreat to which to retire and consequently at the mercy of the enemy.

Vegetius, *Epitoma rei militaris*

The ability to switch from the defensive to the offensive and back again is fundamental if flexibility is to be retained.

The British Military Doctrine[1]

Roman troops on campaign constructed a defended camp at their resting place each night, within which they accommodated, in a carefully detailed and pre-arranged manner, the headquarters, tented lines, animals, followers and baggage of whatever sized formation may have been concerned. Frontinus[2] claimed that they developed the idea after defeating Pyrrhus on the Arusian Plains (*c.* 275 BC), capturing his camp and noting its design. Today, a camp of this nature is referred to by historians as a Roman marching-camp.

Despite the numerous campaigns waged by the Romans on mainland Europe, the remains of only a few of the great number of marching-camp sites which, at one time, must have existed on the continent are still to be found; even those identifiable by crop marks are in scant supply. In Britain, however, particularly England and Scotland, the situation is vastly different. Due to the enthusiasm of eighteenth- and nineteenth-century amateur archaeologists, a survey of Roman camps was commenced at a very early date and, possibly encouraged by this, but certainly stimulated by aerial reconnaissance, the rate of discovery has accelerated in the past forty-five years. Almost three-quarters of the approximately four hundred sites which are known to exist have been recorded during that period.

The marching-camp was an instrument of aggression as much as of defence and it played an essential part, at least down to the third century, in Roman military thinking. It was specifically designed for operations deep in hostile territory. Its standard pattern of layout, which varied according to the size and nature of the forces it accommodated, was important for three reasons. Firstly, and primarily, it offered a secure base from which to continue the advance or, more specifically, the thrust towards conquest; secondly, it provided entrenchments upon which to retire in the event of receiving 'a check in the field'; and, thirdly, the daily construction of the marching-camps left in the wake of the army a series of fortified 'stepping-stones' by means of which the advance could be sustained. Clausewitz, in his day one of the greatest and most original writers on the subject of war, indirectly gave support to the technique. Even under the most favourable circumstances, he wrote,[3]

> and with greatest moral and physical superiority, the aggressor should foresee a possibility of great disaster. He therefore must organise on his lines of operation strong points to which he can retreat with a defeated army. Such are fortresses with fortified camps or simply fortified camps. . . . We must leave behind us a number of troops for the occupation of these strong points as well as the occupation of the most important cities and fortresses. Their number depends on how much we have to be afraid of invasions or of the attitude of the inhabitants.
>
> Napoleon always took great care with these measures for the protection of the rear of his army and, therefore, in his most audacious operations, risked less than was usually apparent.

The concept of the marching-camp fitted comfortably into this strategy, for even when the fortification served no immediate further operational purpose, it still lay available for re-occupation should the military situation require.[4] It was also invaluable in yet another sense. Its layout and construction were ideally contrived to provide a battle drill of operational and administrative behaviour which melded efficiently with the excellent Roman qualities of high military training and discipline.

Clausewitz, when advocating the merits of military theory, observed that it existed so that it should not be necessary to start afresh every time, sorting out the raw material on each occasion and ploughing through it. If theory could be formulated and implanted in the mind of a future commander, he argued, then it would always be ready to hand and in good order. The Romans, with their tactical employment of the marching-camps and other such operational devices, demonstrated that careful planning, with constant rehearsal in the field, leads likewise to a similar but, in this instance, practical end, with every man being aware of what is required of him at every moment. The situation, wrote Caesar, describing a near disaster which confronted him in 57 BC,

> was saved by two things: first, the knowledge and experience of the soldiers, whose training in earlier battles enabled them to decide for themselves what needed doing, without waiting to be told; secondly, . . . the generals did not wait for further orders but on their own responsibility took the measures they thought proper.[5]

*Scene from Trajan's
Column showing tents
within a fortified
enclosure*

Major General J.F.C. Fuller (d. 1966), a fervent apostle of mobile tank warfare, and one whose preaching in this connection was fully justified by the events of the Second World War, regarded the Roman use of the marching-camp with less than enthusiasm. So completely did the spade dominate tactics, he once complained,[6] that the legions seldom accepted battle unless there was an entrenched camp close at hand; indeed, he continued, this occurred to such a degree that not a few Roman campaigns might be described as nothing more than mobile trench warfare. It was his conclusion that the lack of mobility induced by this practice rendered the legionary system unsuited against guerilla warfare. He was, of course, correct in thinking this and there were instances (for example, Viriathus in Spain in 146–140 BC and Scapula's campaign against the Silures in Wales, *c*. AD 50) when the Roman heavy infantry did not perform well in the face of this type of operation. Fuller saw the solution to their difficulties in the provision of a strong, well trained Roman cavalry wing, the many logistical problems for supporting which we examine in detail in these pages.

Undoubtedly, the tribal armies that confronted the Romans during their years of conquest possessed not only the crucial advantage of fighting a war on interior lines, but also a unique knowledge of the topography of their homelands, together with a considerable superiority in numbers. It is thus fair comment to wonder why, with these major benefits and an adaptable tribal command structure, the continental tribes did not turn whole-heartedly to guerilla warfare as a means of defeating their enemy, in the manner that Cassivellaunus used so successfully against Caesar in Britain in 54 BC.[7] The fact remains, however, that they failed to do so and, as a result, the legion remained an effective instrument of conquest. Was Fuller right in his judgment about the consequence of the Roman lack of cavalry? It is difficult to agree with his verdict, particularly when he himself has provided us with good reason to suggest otherwise, by his observation that

> a collateral factor which favoured the expansion of the empire was that, except in Parthia and to a lesser extent in Numidia, the legionaries were never, until toward the end of Rome's supremacy, called upon to face efficient cavalry.[8]

In these circumstances – rarely called upon to face guerilla war, never called upon, until the end of Rome's supremacy, to face efficient cavalry – there is little to indicate that the Roman system of war was incorrectly balanced. Indeed, it may be said that the strategy of the marching-camp, or mobile trench warfare as Fuller so aptly defined it, provided a stability in logistical and military terms which in other ways would have been difficult to contrive.

This, then, in broad terms, was the technique of the marching-camp; let us turn now to the practical application of the concept and its various uses, but particularly as a 'crash' defensive tactic to restore a deteriorating situation and as an instrument of attack.

Two comprehensive descriptions of the construction of marching-camps have been passed down to us, one written with great clarity by Polybius,[9] the other by Vegetius.[10] Both accounts contain valuable information and complement each other. Vegetius underlines the importance of selecting the location of the fortification with care. It should not be overlooked by high ground which might be exploited by an enemy; it should be 'topographically strong'; plenty of wood, forage and water should be accessible; and the 'dimensions of the camps must be determined by the number of troops and quantity of baggage' but, whatever size these may be, they should neatly fit within the perimeter provided for them. To be too small would hamper the defence; to be too large would be unwieldy.

The standard design of a marching-camp, according to Polybius, was in the shape of a square but it had, perforce, to conform to the demands of the ground and to the numbers accommodated. Rectangular layouts are frequently found but, whatever variations are discovered, the sides of the camp are almost invariably straight and the corner angles rounded. Vegetius, the accuracy of whose writings is sometimes queried, relates that

> the camp is formed square, round, triangular or oblong, according to the nature of the ground. For the form of a camp does not constitute its excellence. Those camps, however, are thought to be best where the length is one third more than the depth.[11]

Without doubt, in an emergency and on at least one occasion, Caesar took up a position on a hilltop, locating his baggage in the middle of his defences; but the adoption of circular defensive positions was not routine practice.

A portion of the experimental turf rampart reconstructed in 1966 at The Lunt, Baginton, Warwickshire

The outline of a marching-camp was delineated either by a rampart or a defensive ditch, generally the latter, with a rampart constructed from the spoil, thrown up on to its inside edge. This was then reinforced with sods of earth and strengthened by palisades. The palisades were either fashioned from locally foraged timber or from purpose-made stakes sometimes carried by the soldiery. Both Hyginus and Vegetius quote the dimensions of the ditch as being 5 feet wide and 3 feet deep and, remembering the problems of time and space involved, this would seem to be the maximum possible for a one-night marching-camp. Vegetius, in an unlikely passage, quotes other sizes of trench:

> After the ground is marked out by the proper officers, each century receives a certain number of feet to entrench. They then range their shields and baggage in a circle around their colours and, without other tools than their swords, open a trench nine, eleven or thirteen feet broad. Or, if they greatly apprehend the enemy, they enlarge it to seventeen feet. . . .

It may be that the soldiers cut the initial trace of the entrenchment with their swords, but it is not easy to imagine that thousands of tons of spoil could have been shifted without the use of proper entrenching tools and earth-shifting baskets. We will examine these quantities later, when investigating the time required for the construction of the fortification. According to Josephus,[12] each man was issued with a saw, axe, sickle, chain, rope, spade and basket for entrenching and other such work, although the evidence of Trajan's Column suggests that the last two were not personally carried by the soldiers: they could possibly have been loaded on the 'company' cart.

Polybius, rather more than Vegetius, spells out for us the drills by which the camps were erected:

> Whenever the army on the march draws near the place of encampment, one of the tribunes and those of the centurions who are in turn selected for this duty, go ahead to survey the whole area where the camp is to be placed. They begin by determining the spot where the consul's tent should be pitched . . . and on which side of this space to quarter the legions. Having decided this, they first measure out the area of the *praetorium*, next they draw the straight line along which the tents of the tribunes are set up and then the line parallel to this, which marks the starting point of the encampment area for the troops. . . . All this is done with little loss of time and the marking out is an easy task, since all distances are regulated and are familiar. They then proceed to plant flags . . . [marking out the camp].[13]

There is no difficulty, Vegetius rather obviously remarks, in carrying on the fortification of a camp when no enemy is in sight; but, he adds, in the event of the enemy being close, it was ancient practice that all the cavalry and half the infantry should be drawn up in order of battle, confronting the enemy, with the charge of covering the force working on the entrenchments. The frequently over-confident Julius Caesar on one occasion failed to take this precaution and, as a consequence, nearly paid dearly for his omission.

Roman heavy infantry on the march

The team responsible for the layout of a camp, known as a colour-party in modern military parlance, was mainly composed of ten men detached from each of the centuries of the army on the march. It would have been commanded by a senior tribune or centurion, nominated and briefed by the army commander himself or, possibly, in certain circumstances, by the *quaestor*. Under normal conditions, it moved with the vanguard and carried with it, apart from its own kit, the 'instruments for marking out the camp-site',[14] including a variety of coloured flags and pennants used to identify the various segments of the camp and thus at once attract the attention of incoming troops to their areas so that they might march straight to them without further ado. Since the layout was, in any event, uniform, they would already have had a good idea where to go.

This battle drill, practised by the Romans more than two millennia ago, emphasizes the unchanging nature of a soldier's trade and carries with it a vision of a task which confronted the 7/14th Punjabis with whom the author served in 1942. The battalion had been hastily moved across India, from North-West Frontier Province, to Dimapur, in Assam, with the task of providing cover for General Alexander's Burma Army as it withdrew across the border into the Naga Hills. The battalion was accompanied by its modest complement of sixty regimental mules. When it arrived at Manipur Road Junction, the rail station for Dimapur, it detrained in the teeth of the monsoon which had recently broken, and was then instructed to proceed to Kalemyo, on the Chindwin River. No motor transport was available but the battalion was to make the fullest use of its own pack animals, together with forty Animal Transport (AT) carts which were being provided to lift essential baggage. It was to set forth by route march with all speed. Kalemyo lay 345 miles distant and, since the roads were filled with refugees during daylight hours, movement was to be by night.

Each early afternoon, therefore, the battalion 'colour-party' assembled on the side of the Kohima–Imphal road to hitch a lift. It was armed with 'the instruments for marking out the camp-site', and it also carried with it such essential items as dixies and condensed milk and sugar for the making of hot, sweet tea with which to greet a tired and wet battalion when it arrived in the middle of the night. The flour (*atta*) and oil (*ghee*) for cooking was transported on the AT cart which accompanied each rifle company. The order of march varied daily, for the rear of the column inevitably arrived an hour or more after the leading rifle company was well established.

The task of the colour-party was to select, clear and peg out a camp-site and the layout, as far as the ground would permit, was the same on each occasion. Battalion headquarters, for ease of communication, always lay on the road: transport animals remained with their owners but were to be visited daily by the Animal Transport officer; the first rifle company to arrive at the turn–off point on the approach to the camp moved diagonally deep to the left, the second to the right, and so on. There, liaising with each other, they would dig their defensive positions and construct their two–man bivouacs from a combination of groundsheets and foliage. The problems of preparing food in the rain and in the dark were soon overcome and a drill for marshalling men, animals and transport carts in the jungle and in a different yet broadly identical marching-camp each night quickly emerged.

If our camp-sites on the Imphal road were largely administrative, the writings of ancient historians contain numerous accounts of the manner in which their Roman counterparts were tactically applied.

Tacitus quotes an instance where the concept was used as a means of crash defence in a crisis.[15] The year was AD 14 and Germanicus was in course of returning from a particularly bloody raid across the Rhine when he was ambushed in the Caesian forest by a composite force of German tribes. He was anticipating trouble and was advancing warily behind a screen of cavalry and light infantry, with his main body deployed with 'the first brigade (legion) in the centre, the twenty-first and the fifth on the left and right flanks respectively and the twentieth in the rear'. The remaining auxiliary infantry followed behind. It was an unwieldy formation to adopt in a wood and the enemy took full advantage of the disarray which appears to have followed his initial contact with them. They made a feint against his vanguard and then launched their full force against his rear, causing confusion among the light infantry. Germanicus responded by releasing the XXIst Legion against the enemy, which

> by a single, passionate attack broke through the German army and drove it with heavy losses into open country. Simultaneously, the vanguard emerged from the woods and established a fortified camp. From then on, the journey was without incident. The troops settled into winter quarters, their morale improved. . . .

In this example, the XXIst Legion provided the essential protective cover behind which Germanicus's vanguard was enabled to commence its work.

On another occasion,[16] Caesar manipulated the size of his marching-camp in order to deceive the enemy as to his true strength. When encamped one night, he received information from Cicero that the Gauls were mustering to attack him. Next morning, he broke camp at first light and marched to meet them. He had advanced only 4 miles or so when he saw the Gallic host on the opposite side of a wide valley:

> He felt that he had no cause for anxiety and had better take his time. He therefore halted and made a fortified camp on the most advantageous site he could find. This camp would have been small in any case, since he had no more than 7,000 men and no heavy baggage; but he reduced its size still further by making the camp roads narrower than usual, so that his force would seem contemptibly small.

He then summoned his cavalry to pull back within its fortifications and instructed his soldiery to 'increase the height of the rampart all round and block up the gateways; in doing so, they were to run about as much as possible and pretend to be afraid'. Many of the barricades they erected in this manner were sham and

Scene from Trajan's Column showing legionaries engaged in fort construction. In the foreground the soldiers can be seen digging ditches and transporting the earth in wicker baskets

constructed so as to allow quick exit. As a result, when the enemy drew near with over-confidence, the garrison burst out with unexpected speed and put them to flight inflicting considerable loss of life.

Probably the most illuminating account of the marching-camp as an attacking weapon, although the operation came to near disaster, is to be found in Julius Caesar's *de Bello Gallico*[17] and it deserves detailed attention because of the wealth of guidance it provides, not all of it readily discernible.

The incident occurred in the year 57 BC when, after subduing the Belgae, Caesar had advanced to enter the Nervean homelands, east of the River Scheldt, in central Belgium. He was at the head of an army which comprised 4,000 cavalry and eight legions, together with supporting detachments of Numidian light infantry from Algeria, archers from Crete and slingers from the Balearics. He had met with no immediate resistance and, for this reason, had not yet adopted the tactical order of march he normally practised when moving through hostile territory, namely, with his legions closed up in column of march, under his hand, and with their baggage held back, concentrated under the eye of his rear guard commander. Instead, and undoubtedly because under normal conditions it was administratively easier and more efficient to do so, he was moving with each one of his legions followed closely by its particular baggage train of pack animals and two-horse carts.

Legionary transport, in these circumstances, was frequently burdened with articles of equipment not required at that moment by the soldiery and, since the latter themselves were generally heavily laden, this was helpful. Trajan's Column depicts legionaries marching at ease, carrying shields and javelins, but without helmets. The helmets are shown, together with cavalry shields, loaded on to pack animals and carts following behind the marching troops.[18] When action was about to be joined, these items of kit would have been recovered and exchanged for others of lesser importance in the coming battle, such as shield and helmet covers, for example; but the practice of transporting equipment in this manner, with the risk of the legion's baggage being reverted some distance away to the rear of the column, was uncertain and needed tight control. In 54 BC, when Sabinus's army on the march was ambushed by Gauls, the word was passed along the column for the baggage to be abandoned. The outcome was that 'men everywhere were leaving their units and running to the baggage to look for their most cherished possessions and pull them out, amid a hubbub of shouting and cries'.[19]

The need for soldiers approaching battle to be unencumbered is, indirectly, constantly emphasized by Caesar. In 53 BC,[20] during his operations on the Rhine and after sending the baggage of the entire army to Labienus's camp, he relates how he started out 'for the territory of the Menapii with five legions in *light marching order*'. In 52 BC,[21] while preparing for battle near Bourges with Vercingetorix, he tells us again how he 'ordered his men to *pile their packs* and get their arms ready'. And yet again,[22] in 57 BC he explains how the Gauls planned 'to attack them on the march, when they would be hampered by the presence of transport and dispirited by having to fight with *their packs on their backs*'. This is a subject to which we return below.

As a consequence of Caesar's decision to adopt this routine order of march, his legions would have been separated from one another by a distance of approximately 2½ miles or, in terms of time, a gap of roughly 45 minutes. They would thus have been unable to give their neighbouring formations immediate close support should they have been required to do so. This apparent chink in the Roman general's defensive armour was quickly identified by spies among the many Celtic hangers-on who had attached themselves to his army and they wasted little time in passing news of it to his enemy, the Nervii, already gathering to confront him. The latter saw their advantage at once and quickly made plans to launch an attack against his leading legion, with the purpose of isolating and destroying it.

Meantime, Caesar, who had already been on the march for three days, had learnt from enemy prisoners that the River Sambre lay barely 10 miles from his encampment and that the Nervii, together with their allies the Atrebates, from the neighbourhood of Arras, and the Viromandui, from the Upper Somme, were positioned on its far bank awaiting his arrival. Caesar does not tell us the strength of the enemy ranged against him but it is noteworthy that these three allied tribes had contributed a combined contingent of some seventy thousand men[23] to the large Belgic army he had defeated in battle near Rheims a week or so earlier. The Sambre, at the point where they were assembling, was no more than 3 feet deep and scouts reported enemy cavalry pickets, an arm which the Nervii did not possess in great quantity, posted along the open ground on the opposite side of the river.

There can be no doubt about Caesar's reaction on learning of the presence and obvious intentions of the Nervii and their allies: he welcomed their offer of battle and saw an opportunity to destroy them. His first measure to this end is therefore of interest, for it conveys to us the tactical importance he attached to the marching-camp in the build-up to an attack. He at once despatched a reconnaissance party with orders to discover a suitable camp-site and then pushed forward a cavalry screen, strongly supported by slingers and archers, to provide for their protection and establish contact with the opposition. Then, having set the process in train and, 'in accordance with his usual practice when approaching an enemy',[24] he placed himself at the head of his army and set forth, followed by six legions in close column of march. He was, in his words, unencumbered by heavy baggage which, we may presume, now came under the command of the rear guard, to be watched over by the two legions he had recently raised in Italy.

By this sensible reshuffle of his order of march, Caesar, unintentionally or otherwise, thwarted his enemy's ambition of catching him unprepared; but, at first sight, although events will show this not to be so, he does appear to have exchanged this for yet another problem by divorcing himself from his entrenching tools, unless these were either carried on the man or in 'company' carts retained for the purpose by the forward legions. From the evidence of Trajan's Column,[25] which shows leading troops lightly clad in fighting order and those following behind marching with full kit, then this latter scenario appears to be a possible solution, for they could not have been separated too far from their essential equipment.

Caesar's reconnaissance party had little trouble in finding a suitable location. It selected a high feature, overlooking the Sambre and crowned by a suitable area from where the ground sloped evenly down to the river. Opposite, on the far bank, rose a hill of similar proportions, its summit covered by a wide, heavily shaded wood from which, for a distance of some 300 yards, a stretch of open ground likewise descended a gentle gradient to the river, which separated these two landmarks. In these woods, and doubtless also on the reverse slope behind, stood the enemy host, arrayed for battle but silent and unseen, awaiting the signal to attack.

It was at this moment that Caesar chose to arrive on the scene, followed by his legions. He was in time to watch his cavalry screen, with its supporting light infantry, cross the Sambre, and drive back the enemy cavalry pickets posted along its banks. They then pressed forward up the hill and unsuccessfully began the task of probing the wood at its summit, but refrained from entering because of the uncertainties which lay within its dark interior. Meantime, Caesar's six legions had shed their equipment, stacked their arms and commenced the work of measuring and marking out the camp and constructing its fortifications. Even by his account, they appear to have been left so blatantly unprotected that we can only wonder whether he was guilty of sheer carelessness or supreme arrogance. Either way, he was about to be taught a lesson. Certainly, in later times, he rarely undertook a construction task, for whatever purpose, without first setting in place a strong outpost to protect his legions while they worked:

> The Gauls concealed in the woods . . . were waiting full of confidence. As soon as they caught sight of the head of the baggage train – the moment they had agreed upon for starting the battle – they suddenly dashed out in full force and swooped down on our cavalry, which they easily routed. Then they ran down to the river at such an incredible speed that almost at the same moment they seemed to be at the edge of the wood, in the water and already upon us.[26]

This lively description by Caesar could only have been written by somebody who had experienced the full shock of the assault. It savours of the surprise Zulu attack upon two battalions of the 24th Foot at Isandhlwana in 1879, when Lord Chelmsford, at the head of a 'punitive' expedition, crossed into their territory to apprehend their indomitable tribal chieftain, Cetshwayo. Information of his whereabouts was totally lacking, so Chelmsford, leaving the 24th Regiment to guard his stores and baggage, continued to march forward with his main body in an effort to establish contact.

As Chelmsford's column progressed on its way, the officer commanding the newly organized base despatched a number of mounted scouts, working in pairs, to reconnoitre the countryside surrounding his position. Two of these men noticed a native shepherd boy behaving in a suspicious manner and, as they rode in his direction, he disappeared out of sight, down a hillside, into a valley. Curious to know more, they cantered after him to the edge where they had last seen him. To their horror, the reason for the strangeness of his behaviour at once became apparent. The entire Zulu army of 20,000 men was squatting in the valley, sitting silently, by regiments and in serried ranks, awaiting orders to move. As soon as the scouts came into its vision, the great host rose, uttered a deafening war-cry and then set off in

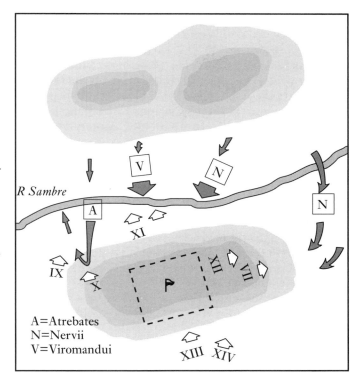

This sketch map illustrates Caesar's use of the marching-camp as an offensive technique. In this instance, during an encounter with the Nervii on the River Sambre, he found himself in difficulties due to over confidence. He was rescued by the arrival of XIII and XIV Legions, the column rear guard, and a robust counter-attack by his experienced X Legion

their pursuit, in silence and with surprising speed. The scouts turned their horses and galloped, *ventre à terre*, to warn their comrades. The latter needed no telling: they had already been alerted by the thunderous noise of feet beating the ground, stirring a great cloud of dust, under which a black tide of warriors flowed towards them, deployed in the traditional Zulu 'buffalo horns' pattern of attack. The technique of this formation was that the 'forehead', comprising the main body of the assaulting *impi*, drove in frontally against the enemy, while the points of the horns carried out an enveloping movement around his flanks. As is now well known, the 24th Regiment, with their ammunition screwed down in boxes for which they had insufficient screwdrivers, were caught entirely unprepared and were annihilated in the engagement which followed.

The Nervii adopted a similar 'buffalo horn' formation as they burst from the woods with their allies to charge across the water and up the hill to Caesar's camp. Caesar's six legions hastily formed rank as best they could but were caught in considerable disarray. The Atrebates, taking the right horn of the attack, had the furthest distance to go and arrived on their objective 'breathless and exhausted with running'. They were also unfortunate to find themselves confronted by the experienced Labienus with the battle-hardened Xth Legion, in the company of the IXth. They were summarily thrown back by the Romans, who pursued them across the Sambre and up the slope to the woods from which they had come.

Here, they made a brief stand before being put to flight. The Viromandui, sharing the centre with a wing of the Nervii, fought their way into the Roman camp before being cast out by the VIIIth and XIth Legions and driven down to the river, where a fierce fight took place in which neither side gained the advantage. The Nervii, with their outflanking drive on the left, enjoyed the greatest success and were on the point of overwhelming the VIIth and XIIth Legions when Caesar arrived and ordered the two legions to combine forces and form a square. This restored confidence to these by now demoralized units and proved to be the turning point of the battle, for help was at hand:

> The two legions (XIIIth and XIVth) which had acted as a guard to the baggage at the rear of the column, having received news of the battle, had quickened their pace and now appeared on the hilltop, where the enemy could see them; and Labienus, who had captured the enemy's camp, and from the high ground on which it stood could see what was going on in ours, sent the Xth legion to the rescue.[27]

This welcome reinforcement restored the balance, so much so 'that even some of the Roman soldiers who had lain down, exhausted by wounds, got up and began to fight again, leaning on their shields'. One can almost hear the cheer which must have greeted the arrival of the Xth!

It is appropriate at this point to consider the factors of time and space which confronted Caesar when undertaking his march to the Sambre, which lay 10 miles away.[28] He had, so he tells us, sent his cavalry 'a little in advance', by which he probably means a distance of half a mile, and he then followed in their wake, marching at the head of six legions. In the event, it is unlikely that he used his entire cavalry wing of 4,500 horse for this purpose: it would have been prudent for him to have retained an element, perhaps 2,000 horse, for his right and left flank guards, with a further 1,000, as army reserve, moving close to his hand at the head of the column. None of these groupings need have moved *within* the column; they therefore play little part in our space calculations, other than to note that the reconnaissance cavalry, together with its attached archers and slingers, would probably have been deployed some 800 yards or more ahead of the main body.

The legionary infantry in this instance moved in close column and, according to Josephus,[29] customarily marched in ranks, six abreast, carrying their shields on their left arm and their javelins in their left hand. Almost certainly many of the soldiers were heavily loaded and it is thus unlikely that there could have been less than a space of 6 feet between the forward toe of one rank to the forward toe of the rank behind,[30] in other words, 3 feet between ranks. There is little more frustrating to an infantryman than to have the foot of a rear file catching his heel as he marches. Thus, a legion 5,000 strong would have occupied 1,667 yards in length, or, in the case of six legions, 10,000 yards (5.68 miles).

In addition, as has already been conjectured, we may expect that carts established with the legions, such as those with each century (total 60) and the *carroballistae* (total 60), with possible additions per legion for ambulance, engineer and other supporting arms (say 30), making an estimated 150 carts per legion or

900 carts with the main body, would also have moved with Caesar's fighting column. If these carts had moved two abreast and each cart required 30 feet of space, then they would have extended a further $450 \times 10 = 4,500$ yards (2.56 miles) or, when taken together with the marching troops, a combined total of 8.24 miles. More accurately, if space were allowed for demarcation gaps between cohorts and legions, this figure might be rounded up to 9 miles; but even this distance, if it were to be prevented from extending further, would have called for strict march discipline. The Romans were well aware of this necessity. Vegetius tells how both their cavalry and infantry were required to practise marching three times a month.[31] The infantry

> were obliged to march completely armed a distance of ten miles from the camp and return, in the most exact order and with the military step which they changed and quickened on some part of the march. . . . Both cavalry and infantry were ordered into difficult and broken terrain and to ascend or descend mountains to prepare them for all kinds of emergencies and familiarise them with the different manoeuvres that the various situations of a country may require.

In the face of the above calculations, if we take into account the ground consumed by the cavalry screen and the party reconnoitring for a camp-site, it is evident that the tail of Caesar's main body would barely have cleared camp by the time his forward elements were approaching the River Sambre. In effect, the heavy baggage, with its legionary escort, would not have moved out of its firm base until such time as the ground between the two fortifications had been secured. Should it so have happened that Caesar had been unable to achieve his objective, or his work preparing the fortifications had been threatened, then there would have been time for the *quaestor* or officer commanding the baggage train to be warned, turn about and re-occupy the whole or part of the fortified marching-camp he had just vacated. If, as is more likely, he chose to re-occupy only a part of it, then he would have needed to adjust the size of the camp to fit his numerical strength.

Caesar records an incident in 52 BC when he failed to take this precaution.[32] He was occupying a temporary camp in the Auvergne mountains, about 4 miles south of Clermont-Ferrand, when he received news that the Aedui were becoming disaffected. Without taking time to reduce the size of the camp for his rear party, he set forth in light marching order, with all his cavalry and four legions, to bring them under control. He left his heavy baggage behind in camp, together with two legions under the command of one of his generals, Gaius Fabius. Caesar dealt swiftly with the Aedui and turned for home. When he was well on his way, he was met by a galloper with the news that

> the enemy had assaulted the camp in full force, and, by continually sending in fresh troops to relieve their tired comrades, had exhausted our men who, on account of the size of the camp, had to remain hard at work on the rampart without relief. Showers of arrows and every other kind of missile had wounded many of the defenders. . . .

Clearly, in the battle on the Sambre, the Nervii had no intention of allowing Caesar's heavy baggage any such avenue of escape. They therefore refrained from attacking him until the head of his baggage train had come into view and, hence, his rear guard was firmly committed to the battlefield.

But let us return to the detail of Caesar's column on the march, for there are other factors of time and space to be considered.

If, as we have calculated, the main body of fighting troops was 9 miles long, then, marching at a generous 3 miles in the hour, the head of their column would have taken more than three hours to arrive at its destination and its tail would not have arrived until three hours later (see Table 3). In other words, a total in excess of six hours would have passed before the last of Caesar's six legions was firmly home or, looking at the timing through the eyes of those responsible for work on the fortifications, the leading legion would have been three hours in the camp before the last of the main body of infantry arrived and became available for work. The rear guard, with the baggage, would have been moving close behind the main body: but we may judge that a half hour would have yet remained before the Nervii, as the Roman commander narrated, 'caught sight of the head of the baggage train' entering the rear gate of the camp. It is a daunting thought that, even at this moment, with an estimated length for the rear guard of 23,625 yards or 13½ miles, some 3½ miles of column would still have been standing ready in its overnight base camp, waiting to move out.

Further, if we regard the army column as a whole, reaching, as it would have done, over a distance of 9 miles (main body) + 13½ miles (rear guard) = 22½ miles, and marching at 3 miles in the hour, then it is evident that Caesar's task of moving his entire force (in this case eight legions) from one camp to another would have taken 3 hours + 7½ hours = 10½ hours marching time, assuming that the two rear guard legions deployed themselves as flank guards to the baggage train they were protecting. Thus, if we add to the marching time a period of one hour for breakfast, packing and saddling-up in the morning, together with a further four to five hours during which to graze the animals upon arrival in the evening,[33] he clearly would have enjoyed a full day's work. Caesar tells us he selected a moment to commence these operations 'when forage began to be plentiful', presumably about mid-May since Gallic farmers knew little about the mixed blessings of fertilizers. In that month of the year and at the latitude of 51° at which he was operating, he might have expected to enjoy some 16 hours of daylight.

The amount of daylight, the fighting strength of the force and the size of its baggage train were factors that controlled the mobility of a Roman army and its rate of advance. If it had been possible for Caesar – and he was far from being alone in this – to make permanent use of river transport for supply purposes and thus reduce his dependency on pack animals, then his logistical problems might have been eased, but his room for manoeuvre would have been reduced. In comparison with his achievement in the Sambre episode,[34] the average distance between Agricola's camps in Scotland (10 to 12½ miles), during the campaign which led to his victory at Mons Graupius, are remarkably little different, even though, through the medium of the Roman navy's *Classis Britannica*, the latter may have received some provisioning by sea. In this respect, it is also notable that

TABLE 3
COLUMN ON THE MARCH
March timings

Data: a. March rate – 3 miles in an hour
b. Distance to Camp II – 10 miles
c. Overall length of Marching Column – 22.5 miles
d. Departure time – H hour

Serial No.	Time	Event	Remarks
1.	H hour	Recce Group departs Camp I	–
2.	+ 0 h 10 m	Vanguard departs Camp I	Followed by Command Group and Main Body
3.	+ 3 h 20 m	Recce Group arrives Camp II	–
4.	+ 3 h 30 m	Vanguard arrives Camp II	Followed by Command Group and head of Main Body
5.	+ 3 h 30 m	Camp layout commenced	Tail of Main Body departs Camp I
6.	+ 3 h 30 m	Head of Baggage train departs Camp I	
7.	+ 4 h	Protective Screen deployed	After arrival of first legion
8.	+ 4 h 30 m	Fortifications commenced	After arrival of second legion
9.	+ 6 h 30 m	Tail of Main Body arrives Camp II	–
10.	+ 7 h	Head of Baggage train arrives Camp II	Slightly slower march rate than Main Body
11.	+ 7 h 30 m	Fortifications complete	–
12.	+ 12 h	Tail of Baggage train arrives	Column complete at Camp II

(Source: Caesar, *de Bello Gallico*, II, 17–28)

Agricola's army of an estimated 5,000 cavalry and 16,000 infantry was considerably smaller in size than Caesar's eight legions in Gaul.

It has already been mentioned that two prime factors to be considered when selecting the site of a marching-camp, apart from its defensive qualities and its ability to be adapted to the Roman layout requirements, would have been the availability of water for man and beast, as well as an adequacy of grass for grazing and bedding. The very numbers concerned (in broad terms, in Caesar's case, 45,000 men and 16,000 animals) serve to emphasize the scale of the problem and underline the fact that the presence of a strong-flowing stream or river, providing easy access from the banks, would have been essential. This points, yet again, to the advantage of a riverine supply route. In the case of grazing, the imagination cannot grasp the confusion which would be created by turning such numbers of animals out to grass, even by day, unless tight control were exerted. Vegetius' comment, that 'the darkness of night, the necessity of sleep and the dispersion of horses at pasture afford opportunities of surprise'[35] is, almost certainly, not intended to suggest that animals were permitted to graze after dark but serves to understate the whole, operationally vital, problem of feeding the animals.

Today it may be expected that, on such an occasion, horses would graze on a scale of twelve to an acre; but modern animals are fussy eaters, picking out the succulent bits of grass and passing over the more unappetizing patches. The Roman cavalry horse and pack animal were less highly bred and, as a consequence, less particular in their eating habits. If, in their case, we allow a total of twenty grazing animals per acre, this poses a requirement of 800 acres (or 1¼ square miles) of reasonable quality grass, dependent, of course, upon the time of the year. The hay crop in Gaul before mid–April would have been scant and of poor nutritional value, demanding a yet wider grazing area. Indeed, the lack of an adequate supply of grass frequently caused a delayed start to a campaigning season. Thus, with this large area of pasture land to be watched over whenever the animals were put out to graze, the army commander was presented with a considerable security problem. If Vegetius tells us correctly, the burden of resolving it fell upon the cavalry:

> The cavalry furnish the grand guards at night and the outposts by day. They are relieved every morning and afternoon because of the fatigue imposed by this duty upon the men and horses. It is particularly incumbent upon the general to provide for the protection of the pastures and of the convoys of grain and other provisions either in camp or garrison, and to secure wood, water and forage against the incursions of the enemy.[36]

Purely to illustrate the scale of need, if a minimum stabling space of 12 feet × 3 feet were allowed for each of the horses/mules mentioned in the example above, and all the animals were stabled tightly together, each within this space, there would have been a total requirement of 64,000 square yards or 13.2 acres (5.35 hectares), without taking account of an unknown number of pack-transport reserves or remounts for the cavalry. Patently, it was administratively a formidably large but, in the military practice of the day, not unusual total of animals. In camp they would have been accommodated in the areas of the units responsible for them.[37] The pack animals provided for each of Caesar's 4,800 *contubernia*, for example, are thought to have stood in front of the tents of the 'sections' to which they were attached. The servants in whose charge they lay probably set down their beds in a corner of the section tent, keeping with them the saddlery and stores for which they were responsible. The sub-unit cart would probably have been parked, off the road but accessible for use, at the end of each line, which, in the case of the Roman infantry, housed a maniple or double century.

The cavalry equivalent of a century, a *turma*, comprised thirty-two troopers, and was commanded by a *decurion*. Their horses were probably tethered face to face, in

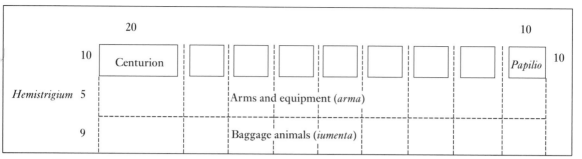

The arrangement of the century's tents within the marching-camp, according to Hyginus

the enduring military fashion, secured to a central picket line over a length of some 60 feet. This distance would have fronted four tents for the trooper and a double sized tent for the *decurion*. It is likely that a further tent would have been provided for servants, in which saddlery and spare stores could have been housed. The legionary, as opposed to auxiliary, cavalry were probably provided with a servant per man. The scale of provision of servants for auxiliaries is uncertain. They possibly rated, like the infantry, one for each eight men, together with a mule for tent and other camp equipment.

The scale of the old marching-camp at Rey Cross, lying between Scotch Corner and Carlisle on the windswept summit of the Stainmore Pass, permits us to consider a practical example. It is considered to have accommodated the ten cohorts of a legion,[38] together with a complement of auxiliaries, partly cavalry. It is thus a microcosm of Caesar's army and the 18–20 acres it occupies may, therefore, be fairly used to deduce the size of the marching-camp of eight legions set up by him on the Sambre, that is some 160 acres × 4,840 square yards (per acre) = 774,400 square yards. By taking the square root of this total, we see that his camp would have been half a mile, or 880 yards, square in size. From this we may further deduce that the defensive ditch of the camp, 5 feet wide and 3 feet deep, would have extended an overall distance of 4 × 880 yards = 3,520 yards (10,560 feet) and the spoil from it would have totalled 2241.4 cubic metres. Dependent upon the nature of the soil, one man could shift 0.4 to 0.7 cubic metres per hour, including a throw not exceeding two metres (see Appendix 3).

On this basis, the task would have taken 3,202 to 5,603 man-hours of work, depending upon the condition of the soil and, allowing one man for every 5 feet of working space, it would have required a labour force of some 2,112: more than this number would have caused overcrowding. Naturally, other fatigue parties would also have been involved, working on the ramparts and standing by to provide relief. Thus, presuming its line had already been pegged out by the advance party, the construction of the trench would have taken between 1.52 to 2.65 hours of work. In reality, we are probably talking of a period of 2¼ to 3½ hours' work after the arrival of the main body. Plainly, the task of digging would have been considerable and, however much the soldiery may have achieved with

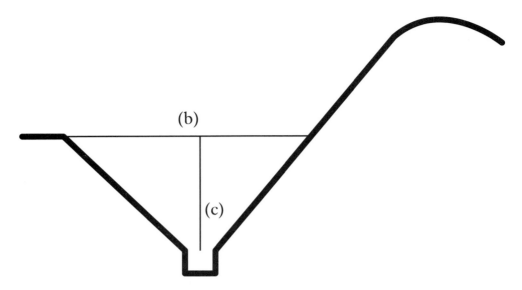

Source	Trench Size Length (in feet)		Man-hours Day Night		Approx. tonnage
1. Hyginus	5 (b)	10,560	(d) 5,602	8,403	3,361
	3 (c)		(e) 3,201	4,801	
2. Vegetius	9 (b)	10,560	(d) 23,534	35,301	14,120
	7 (c)		(e) 13,448	20,172	

Notes: a. Night-time work load, by light of full moon, calculated at two-thirds of daytime. b. Width of trench in feet. c. Depth of trench in feet. d. 2,241 cubic metres (.0283 cu. metres = 1 cu. foot) at 0.4 cubic metre per hour on difficult earth surface. e. 2,241 cubic metres at 0.7 cubic metres on easier ground. f. 9,413 cubic metres at 0.4 cubic metres per hour. g. 9,413 cubic metres at 0.7 cubic metres per hour.

Marching-camp defences (source: Royal Engineers Pocket Book, Labour Figures for Common Engineering Tasks, Table 112, p. 14)

their swords as Vegetius has suggested, it would have been imperative to carry the entrenching equipment and soil-shifting baskets well forward in the column, so that the exercise might be commenced and completed with all haste.

Equally, it is clear that before work could start a strong defensive screen would have had to be in position. In normal circumstances, the first legion to arrive would have been detailed for this duty, thus delaying the start of work until the arrival of the next in line. This probably explains the gamble Caesar took on the Sambre by failing to put a protective screen of infantry in place at the first opportunity.

The rampart of a marching-camp was constructed so as to be a distance of 200 feet from the lines of tents.[39] The space thus created had a multitude of uses.

Most importantly, it placed the tented accommodation out of range of any fire-arrows or javelins which might be fired or cast into the camp. It provided a highway by which the occupants of the camp might gain access to their quarters without committing the military sin of passing through another unit's lines. It was also employed as a parade-ground, a collecting point for plunder brought into the camp and for cattle rounded up for use by the commissariat. The plunder would have attracted traders like bees to a honeypot. At Atuatuca, it will be remembered,[40] the German cavalry attack on Cicero's camp overwhelmed the traders 'who had their tents at the foot of the rampart and left them no time to get away'. Here also, we would have found the *carroballistae* of the entire force, brigaded as one artillery whole, to add formidable strength to the defence of the ramparts.

The marching-camp, it has been said,[41] provided an important form of psychological security for Rome's troops but it could not keep a determined foe out of Europe: but to say this is not fully to recognize its concept. The marching-camp was an instrument of systematic attack and the highly trained and disciplined legions that provided its garrisons had little need of a psychological boost. The marching-camp technique, allied to their supreme confidence in themselves, furnished the legions with a formula for conquest which carried them to the furthest corners of the Mediterranean basin and beyond. Victory in war, Vegetius advised the emperor Valentinian,

> does not depend entirely upon numbers or mere courage; only discipline and skill will ensure it. We find that the Romans owed the conquest of the world to no other cause than continual military training, exact observance of discipline in their camps and unwearied cultivation of the other arts of war.[42]

As the Roman appetite for conquest waned, so the concept of the marching-camp was abandoned and the empire settled down to the defence of its frontiers. Unfortunately for Rome, she had no defence strategy of comparable genius to set in its place.

CHAPTER FIVE

Supporting Arms and Weaponry

And he [David] took his staff in his hand and chose him five smooth stones out of the brook and put them in a shepherd's bag which he had, even in a scrip; and his sling was in his hand; and he drew near to the Philistine.

I Samuel, 17: 40

But Mechanics, by means of one of its smallest branches – I mean, of course, the one dealing with what is called artillery construction – has surpassed argumentative training on this score and taught mankind how to live a tranquil life. With its aid, men will never be disturbed in time of peace by the onslaughts of enemies at home and abroad. . . .

Heron of Alexandria (second century)

In land warfare, terrain, at all levels of operations, is a critical factor, principally in the manner by which it provides cover for an assault, commands a position, presents obstacles to an enemy's advance or, alternatively, provides obstacles within which a defensive position may be adopted. Additionally, as we have discussed in earlier chapters, there is the importance of the administrative application of the terrain to military supply needs. A careful study of terrain and the areas of ground vital to the successful achievement of his military objectives forms part, therefore, of every commander's operational preparations and is a principle as old as warfare itself. Vegetius,[1] in his treatise on the art of war which he dedicated to his emperor, Valentinian II, made this point with clarity. Good generals, he advised,

are acutely aware that victory depends much upon the nature of the field of battle. When you intend to engage, endeavour to draw the chief advantage from your position. The highest ground is reckoned the best. Weapons thrown from a height strike with greater force; and the party above their antagonists can repulse and bear them down with greater impetuosity, whilst they who struggle with the ascent have both the ground and the enemy to contend with. . . . If you depend on your foot against the enemy's horse, you must choose a rough, unequal and mountainous situation.

Vegetius thus saw three advantages in the infantry axiom of 'going for the high ground': it impeded the enemy cavalry, it reduced the momentum of the enemy assault and it improved the weight of the available firepower, an important consideration, as much in attack as defence.

We do not need to remind ourselves, for many of the world's earliest archaeological discoveries are weapons, that man, habitually, is a quarrelsome animal, driven along this path as much by a determination to survive as by a natural aggression. Many historic finds, such as flint daggers, axe-heads attached to handles by thongs, spears of a wide variety, and bows with flint-headed arrows, the most ancient of which have lain concealed for 10,000 years and more, bear testimony to this fact. They also provide evidence of mankind's unceasing quest for improved 'firepower', from the moment that he instinctively cast the first stone or picked up a piece of flint to fashion as an arrow-head, until his manufacture of the devastating nuclear weapons which lie today in many of the world's arsenals. In August 1945, when atom bombs were dropped with devastating effect upon the Japanese cities of Hiroshima and Nagasaki, a war-weary world saw them as the ultimate weapon, which would guarantee peace for future generations. Heron of Alexandria, as the implications of the new torsion artillery weapon increasingly revealed themselves to his generation, was persuaded to take a similar view of this powerful new piece of artillery. With its aid, he wrote,

> men will never be disturbed in time of peace by the onslaughts of enemies at home and abroad, nor, when war is upon them, will they ever be disturbed. After peace has continued for a long time, one would expect more to follow when men concern themselves with the artillery section:[2] they will remain tranquil in their consciousness of security, whilst potential aggressors, observing their study of the subject, will not attack.

Heron must have been deeply disappointed by the subsequent history of events, just as we, in our time, have had our hopes and expectations violently dashed by a deeply troubled post-Cold War world, wracked with civil war and strife.

In modern practice, considerable firepower is brought to bear by the infantry itself, augmented, as the situation may demand, by artillery and other weapons of differing calibres and diverse means of propulsion. In Roman times, on the other hand, firepower, provided by the simplest of weaponry such as bows and arrows and slingstones, was concentrated in special units of light-armed troops to provide support for infantrymen, who themselves were each armed solely with a sword, perhaps a spear and a handful of javelins, and carried a shield for their protection. The javelins provided a form of unit firepower and were employed, both in defence and in the set-piece battle, to disrupt an enemy assault before it reached the crucial stage of hand-to-hand fighting. The general design of the javelin bore close resemblance to the carefully honed weapon used today for hunting and tribal warfare by the Karamajong and Turkhana peoples, whose territories adjoin each other in northern Uganda and Kenya, south of the Sudan border. I have seen these cast, with varying skill but without great difficulty, over distances ranging between 100 and 150 feet.

In each case, ancient or modern, the guiding principle for the use of firepower remains unchanged: concentration of fire is essential if maximum effect is to be obtained. By this means, according to the circumstance and whatever the range in relation to the stage of development of the weapon, an enemy's power of manoeuvre may be hampered, his forces neutralized and his will and ability to fight may be destroyed.

Initially, the three main infantry support weapons were the bow, the javelin and the sling, not necessarily in that order. The origin of the first two essentially belonged to Asia. The sling, shrewdly employed by David in his fight with the Philistine, Goliath, was widespread as a weapon, popular because of its simplicity, its lightness and its ease of manufacture. It comprised a piece of leather to house the missile, and two thongs, one of which was secured to the throwing hand and the other held, simultaneously, between the thumb and forefinger of the same hand. It was then cast, after a single twirl round the head, the missile being 'fired' at the moment that the second thong was released, its range being related to the angle of discharge, the length of the thongs and the amount of energy imparted by the thrower. David kept his supply of ammunition in a shepherd's bag, slung across his shoulders, thus keeping both hands free for the task of making his cast and then reloading. A slinger in a battle illustration on Trajan's Column, on the other hand, is shown carrying his reserve of stones in the folds of his cloak, draped over his left arm.[3] This appears a clumsy arrangement, which must have restricted the mobility of the soldier and slowed his rate of fire.

The use of the sling is widely believed to have been perfected in the Balearic Islands of Majorca and Minorca. Vegetius relates[4] that the inhabitants of these islands were taught the art of sling-stoning from a very early age, when, as children, they were not allowed food by their mothers till they had first made an accurate cast of a stone at a target. Islanders from Rhodes, on the other hand, also possessed a worthy reputation for their skill with the weapon, partly gained during the siege of Syracuse in the early third century BC. The development of leaden bullets as slingshot, shaped like acorns and known as *glandes*, has been attributed to these people.[5]

Lead missiles would have had many benefits, both logistical and military, for they would have been smaller than a stone, weight for weight, and thus larger quantities could have been carried both on the man and within the re-supply organization. Moreover, since they were fabricated, their availability would have saved time wasted in a search for the right size and shape of replacement stones, if indeed any were to be found at all on certain types of terrain. The provision of lead bullets introduced a logistical problem but this was probably no more taxing than that presented by the re-supply of stones, which generally were either rounded artificially or manufactured in baked clay to the required shape, weight and size. Livy underlined the need for the careful selection of missiles in his account of an encounter between Romans and Gauls in 189 BC:

The Gauls had insufficient protection from their shields, which were long, but not wide enough for the size of their bodies, and, besides that, were flat in surface. Furthermore, they had no other weapons than swords, which were of no service to them on this occasion, since the enemy did not engage them in close combat. The missiles they used were stones, but not of convenient size, since they had not collected them in advance, but every man took what came to hand in his agitated search; and they used them like men unused to the work, with neither the skill nor the strength to lend force to their impact.

All of which emphasizes the importance, well understood by generations of soldiers, of the dual values of victory, for this does not result solely in the defeat of the enemy but gains for the victor the retention of the battlefield, with its reward of a plethora of military salvage, whether comprised of sling-stone missiles in ancient days or armoured tanks in more modern times. A sensible commander, as we are reminded by Livy, seizes every opportunity to recoup such losses. When assault troops halted to gain breath, he tells us, during their ascent to attack an enemy fortification in the Galatian mountains, the legions 'made energetic use of the interval in collecting weapons all over the hills to ensure a sufficient supply of missiles'.[6]

During his early months under recruit training, every Roman soldier was taught the art of throwing stones, both with the hand and with the sling, the stone thrown by hand weighing about 1lb.[7] To this end, archers and slingers alike, the latter utilizing a device named by Vegetius as a *fustibalus*,[8] were required to set up aiming marks of bundles of twigs or straw and to strike them at a distance of 600 feet. A *fustibalus* is a rarely mentioned small-arm of olden times, defined by *Freund's Latin Dictionary*[9] as 'a sling-staff, an offensive weapon consisting of a staff with a sling attached'. The extra length and rigidity of a staff, fitted in this manner with a sling, would undoubtedly have provided considerably more distance than the range achieved by the solely hand-operated version, and one is driven to consider whether this was the purpose of the staff carried by David, mentioned in I Samuel 17:

40. . . . He took his staff in his hand and chose him five smooth stones out of the brook, and put them in a shepherd's bag that he had, even in a scrip; and his sling was in his hand; and he drew near to the Philistine.
41. And the Philistine came on and drew near unto David; and the man that bear the shield went before him. . . .
43. And the Philistine said unto David, Am I a dog, that thou comest to me with staves?

The role of the sling-staff contingent would have been similar to that of the archers, namely, to engage advancing enemy at long-range and break up the momentum of their advance. As the enemy drew near, the *fustibalatores* would have fallen back, through the ranks of the massed infantry, to assume a position at the head of the reserves. Conventional slingers, also, were sometimes used in this manner but, during the set-piece battle, they were normally posted, frequently intermingled with archers, on the left or right flanks of the main body as protection against cavalry attack. In an assault upon a fortified objective, their task would have been to provide covering fire for their own troops, by so directing it as to neutralize enemy counter-fire. In an advance, slingers were frequently employed as light infantry, probing ahead of the main body. Caesar, for example, when seeking out a Remi stronghold in 57 BC, records that he pushed his cavalry across the River Sambre, with slingers and archers, to engage and drive off the enemy's horsemen.[10]

The author[11] of *de Bello Africo* records a bizarre use of slingers against elephants during Caesar's operations at Ruspina against Labienus. These huge animals were generally employed to break into the main body of infantry, thus simplifying the work of the assaulting troops. The counter-stroke, by those under attack, was to bring the slingers forward and discourage the advance of the elephants by pounding them with slingshot. But, the chronicler explains rather ungenerously, the animals, for all their considerable battle training, were uncouth creatures, equally dangerous to both sides. The not unreasonable reaction of the elephants to this unwelcome attention was to turn aside and race back upon their own troops, carrying confusion into their ranks. Scipio, the senior cavalry general opposing Caesar in North Africa, undertook to stop this undesirable habit, but it proved a 'difficult and slow process' of hard, repetitive training. Nevertheless, it worked in this manner: Scipio drew up three lines to simulate a battle situation,[12]

. . . one line of slingers, facing the elephants, to take the place of the enemy and to discharge small stones against the opposing front formed by the elephants; next he arranged the elephants in line, and behind them drew up his own line so that, when the enemy proceeded to sling their stones and the elephants in their consequent panic wheeled round upon their own side, his men should receive them with a volley of stones and so make them wheel around again . . . in the direction of the enemy. . . .

The years between the death of Alexander the Great (323 BC) until the ignominious defeat of Antiochus by the Romans at Magnesia (190 BC), constitute the prime period during which elephants played an important part in western warfare. Without doubt, they were used after this, for they were employed, among other occasions, by Marcius Philippus on operations in Macedonia in 169 BC, by both Caesar and Labienus in 46 BC in North Africa and by the emperor Claudius, who took a few with him from Gaul to Britain in AD 43. Their presence at this moment probably had more to do with the emperor's desire to impress the inhabitants with his power and importance than with any military purpose. Elephants, however, were never used again in the numbers which had been common in the early days of the Seleucid dynasty (312–64 BC), when Seleucus I defeated his rival, Antigonus, at the battle of Ipsus in 301 BC and was able to put 480 animals into the field.

Seleucus, once one of Alexander the Great's generals, had commanded the Macedonian infantry against King Porus of India in 326 BC, in battle on the Hydaspes river. Porus possessed a force of 200 elephants and was only defeated after an exhausting and bloody struggle. As a result of this experience, Seleucus at once determined that this was an arm of war to be obtained at almost any price and, later, in 305 BC, when his efforts to expand his Babylonian Empire to the Indus were halted by Candragupta on the eastern frontiers of Persia, he agreed territorial concessions in exchange for 500 elephants. Doubtless these were the animals that won him the day at Ipsus.

It thus happened that the animals used by the Macedonian Powers were of Indian stock and those employed by Carthage were African. The African elephant, the largest living land animal, stands on average 10 to 13 feet at the shoulder and weighs up to 8¼ tons; the Indian species is roughly 33 per cent smaller, reaching only to 10 feet at shoulder height and weighing some 5½ tons. Elephants have a voracious appetite and spend many hours eating. They may consume more than 225 kg of grasses and other vegetation in a day; unless the terrain was suitable, therefore, they could have been a logistical liability if employed in large numbers. Thus, the 500 animals employed by Seleucus would have required a daily diet of roughly 110 tons of fodder. In November 1944, during the Burma campaign, the Indian battalion with which I was serving was thrusting southwards up the Myittha Valley, with our forward patrols clearing the west bank of the Chindwin. During this probing period, we intercepted a valuable haul of 20 elephants, together with their handlers (mahouts), belonging to the Irrawady Steamship Company. They had been summoned to join the Japanese to assist with their transport problems. On arrival with us the animals were picketed around our position and, during the night, with their fore legs shackled by chains, they were allowed to forage in the jungle for fodder. Sleep did not come easily to us as they pushed their way through the undergrowth, snorting and rumbling, rattling their chains, and uprooting young trees in a quest for fodder. It is unlikely that the Seleucus at Ipsis permitted their large elephant corps a similar licence to feed in this dangerous, haphazard manner.

The elephant as a military weapon appears to have been encountered by the Romans first in Lucania, southern Italy, in war with King Pyrrhus of Epirus, when he twice defeated Roman legions in hard fought battles; but, although a novel military device, their use soon lost terror for experienced troops and the Romans, by opening up wide corridors in their ranks of infantry and so enabling the latter to isolate and encircle the animals, quickly learnt how to handle this form of attack. Scipio, once again, was the author of this tactic. He employed it for the first time in 202 BC, in a battle against Hannibal at Zama in North Africa. The front line of his infantry was, in the Roman custom, formed by maniples of *hastati*, which were habitually drawn up in battle with intervening gaps of similar dimensions between each maniple. Behind the *hastati*, the *principes* were arrayed, in a second line, not, on this occasion, covering the gaps between the front line maniples, which was the normal drill, but lined up behind the maniples in front. The *triarii* of the third line behaved in a similar manner. Scipio filled the intervals between the front line maniples with *velites*, lightly equipped skirmishers, who had instructions, in the event of an elephant assault, to conduct a fighting withdrawal down the open corridors thus formed behind them and, if overtaken, to filter away to left and right to seek shelter among the flanking heavy infantry. The victory achieved by the Romans at Zama provided them with a stepping-stone towards their final defeat of the Carthaginians.

The presence of these huge, ungainly animals in an army must inevitably have applied a brake to its mobility, not by their speed of march, it must be added, but by their sheer size; and the military benefits they brought would necessarily have had to be weighed against this limitation. This fact was clearly demonstrated in an incident experienced by Hannibal during the Second Punic War when, with a force of 50,000 infantry, 9,000 cavalry and 37 elephants he had surmounted the Pyrenees and was now marching through southern Gaul to Italy. Scipio, having learnt that his enemy was in course of crossing the River Rhône, set out on a forced march to intercept him. Hannibal, however, who had encountered a little local opposition, had by this time managed to get the bulk of his army to the left bank of the river, with the exception of his elephants and their guardians. He countered Scipio's move by despatching the whole of his cavalry force southwards to cover his movements and ordered his infantry to follow them in route march. Meantime he remained behind to supervise the passage of his animals.

He now constructed a series of solidly built rafts which, when lashed together, were 50 feet in overall width. The first pair of rafts rested wholly on dry land; others were then attached on the far side and fed into the waters of the river, until they projected a distance of some 200 feet. At this point, two more equally well constructed rafts to transport the elephants were firmly fastened to each other and placed in the river, but they were so connected to the main pontoon that the lashings could easily be cut. The whole pier was now piled up with quantities of earth until its surface had been brought up to the level of the bank and the elephants, which had previously resisted all efforts to get them afloat, were led along the earthen causeway, with two female animals leading the way:

> As soon as they were standing on the last rafts, the ropes holding these were cut, the boats took up the strain of the tow ropes and the rafts with the elephants standing on them were rapidly pulled away from the causeway. At this the animals panicked and at first turned round and began to move about in all directions but, as they were by then surrounded on all sides by the stream, their fear compelled them to stay quiet. In this way, and by continuing to attach fresh rafts to the end of the pontoon, they managed to get most of the animals over, but some became so terror-stricken that they leapt into the water when they were halfway across. The drivers of these were all drowned but the elephants were saved, because through the power and length of their trunks they were able to keep these above the surface and breathe through them, and also spout out any water which had entered their mouths. In this way most of them survived and crossed the river on their feet.[13]

It has been argued that the military role of the elephant in antiquity can be equated with that of the tank in the twentieth century.[14] Without doubt, it is possible to see the force of this suggestion as far as the earliest armoured vehicles are concerned, introduced in 1917 and used with such outstanding success at Cambrai as a device to break through the rain-sodden and blood-soaked German Hindenberg Line. Their dramatic arrival on the battlefield caused the enemy to scramble from their dugouts and shelters, in shocked amazement at their 'grotesque and terrifying' appearance.[15] The Roman soldiery, in their first encounters with elephants, behaved no differently.

A war elephant, surmounted by a combat tower, used by Hannibal in the Punic Wars. Elephants were often described as the tanks of ancient warfare

Scipio provided further grounds for comparison between animal and tank in an engagement with Caesar at Sidi Messaoud, midway between Ruspina and Leptis, when he arrayed his cavalry for battle, interspersed with elephants equipped with armour and towers,[16] the latter manned with archers. His worthy opponent at once recognized the manner in which the size and number of these animals 'gripped the minds of his soldiers' and appreciated the importance of having a similar strike force of his own. He promptly ordered a number of animals to be brought across from Italy, to enable his troops to familiarize themselves with this intimidating weapon and to train with them. Within a few weeks, he was threatening the town of Thapsus with his own army, supported by a contingent of sixty-four elephants similarly armed and caparisoned.

The elephant had three limited uses in ancient warfare: to act as a screen against enemy cavalry; to attack and penetrate the infantry mass; and to break into a fortified position; but, whenever employed in this last function it was generally unsuccessful. Moreover, when brought under fire, its behaviour, in large numbers, was unpredictable and frequently as dangerous to friend as to foe. Caesar witnessed evidence of this at Thapsus when, in a spontaneous assault on the enemy, his right wing, made up of massed slingers and archers, directed a rapid fire of missiles against the elephants covering Scipio's Moorish cavalry. The beasts, 'terrified by the whizzing sound of the slings and by the stones and leaden bullets directed against them, speedily wheeled round and trampled under foot the massed and serried ranks of their own supporting troops behind them. . .'.[17]

The Romans derived many of their infantry tactics, much of their weaponry and certainly a great deal of their knowledge of arrow-shooting and stone-shot artillery weapons from the Greeks. The latter, in their turn, had taken the Assyrian foot-soldier as a model upon which to base the Grecian army hoplite, their heavily armed spearman, who was destined to gain for himself a high reputation in military history. The Assyrians were also specialists in the art of siege warfare and their expertise in this was passed, through bitter experience, to Syria and Phoenicia and, ultimately to Carthage. From here, once again, military expertise found its way to Greece and thence to Rome. The Assyrian Empire, like all 'earth's proud empires', passed away but many of their peoples, in particular the Hamii, Ituraei and Damasceni, continued to provide mercenary soldiers to any who chose to buy their swords and many, in due course, served the Romans as auxiliaries alongside other specialist contingents provided by Rome's client states.

Tribal levies, possessing a wide variety of specialist skills missing from the Roman military arsenal, fought alongside legions in most theatres of operations and proved an invaluable complement to the Roman military establishment. They were categorized as auxiliaries. Evidence of the presence of Hamian archers has been discovered on both frontier walls in northern Britain.[18] Julius Caesar, as we have seen, was supported by Celtic cavalry, archers from Crete, slingers from the Balearic Islands, Numidians from Algeria and many others. The Numidians could boast proficiency in the use of both these supporting weapons. Vespasian (AD 69–79), during his wars against the Jews, employed Arabian archers and Syrian slingers. Probably the most colourful army of the era was brought together in North Africa by Labienus, when engaged in civil war against his erstwhile friend, Julius Caesar. Labienus's polygot force comprised 120 elephants, 1,600 Gallic and German cavalry, a further cavalry contingent of similar size commanded by Petreius, 8,000 Numidians who rode without bridles (a useful skill for mounted archers), a body of some 6,500 infantry recruited from half-castes, freedmen and slaves, and the customary contingents of provincial light infantry, composed of archers and slingers.[19]

The examples are countless. A segment of Trajan's Column, which features the *auxilia* in action, underlines the wide variety of such support troops.[20] They range from dismounted cavalrymen of obscure nationality, armed with long spears and carrying oval shields marked with a star and crescent, to a half naked club man, carrying a knobbed truncheon, a long slashing-sword and a shield for his protection. It is possible he came from the Aesti, a German tribe of whom Tacitus wrote that they seldom used 'weapons of iron, but clubs very often'.[21] Further to his right, a slinger is depicted in the act of casting a stone, while upon the opposite flank of the attacking column a company of archers is illustrated, pressing forward to the left. These men, depicted with high cheek bones and aquiline noses, appear to be from the Middle East, are possibly Syrians, and are shown wearing leather jerkins and voluminous skirts. They carry composite bows in their left hands and, on their backs, quivers of arrows, so positioned as to be conveniently reached by their free hand. The bows appear to be of the light but powerful 'Turkish' pattern,[22] a derivative of the Oriental composite bow, 3 ft 9 in in length when measured along its outer curve and 3 ft 2 in when fitted with a bowstring of 2 ft 11 in. The war arrow it discharged measured 2 ft 4½ in in length and required a draw weight of 118 lb to pull the bowstring back to its full capacity.

Composite bows were the outcome of necessity, for they originated from a time and place when the natural materials for bow-making were unobtainable. Their use appears to have been largely confined to such areas as modern-day Russia and China, from whence they found their way to India and the Middle East, the northern part of North America and the coastal districts of Greenland[23] – those areas which at one time endured the Ice Age. Bows constructed in this manner have been discovered in ancient Egyptian tombs.[24] Where it was available, the core of the composite bow was made of wood. On the belly side of the bow, that is to say the surface facing the archer, horn was glued; on its opposite side, sinew, taken from the neck tendon of either a stag or an ox, was affixed in a similar manner. Thus, when the bow was drawn, the sinew stretched. When the bowstring was released, the bow, pulled sharply by the sinew as it shortened to its normal length, snapped back to resume its original shape.

It can be imagined that stringing a powerful reflex weapon of such force presented problems to a man of average strength. Sir Ralph Payne-Gallwey, who made a lengthy study of Oriental bows, commented that he had heard of no one of

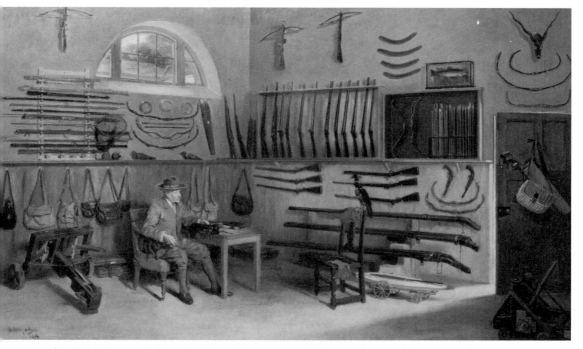

Sir Ralph Payne-Gallwey in his workshop surrounded by crossbows and Roman artillery weapons

his day who could string a Turkish bow unless by mechanical means, 'yet formerly the Turkish archer, unaided, could do so with ease', by bending it between his legs while stooping down to fit the string. He judged that the composite bow, without difficulty, would have been capable of driving a 1 oz war arrow a distance of 360 to 400 yards, well in excess of the performance of the European longbow with a comparable range of 230 to 250 yards. These distances are dwarfed, however, by those achieved by light target arrows shot from a Turkish-style weapon, where ranges in excess of 800 yards have been registered.[25]

The re-supply of arrows in battle was seemingly not considered important until the middle of the first century BC. The reason is not far to seek, for it lay in the very nature of battle itself, which was defensive and confrontational until, in the last resort, the swordsmen and spearmen were released for hand-to-hand fighting. Even Rome's wars of conquest have been described as 'mobile trench warfare'. Caesar, although he employed archers in his tactical armoury, rarely used them in the large numbers which became the custom in later years. In North Africa, at Ruspina, he mustered only 150 bowmen,[26] alongside thirty cohorts. Some weeks later, at Cercina, he received reinforcements from Italy which included the XIIIth and XIVth Legions, 800 Gallic cavalry and 1,000 slingers and archers,[27] and for many years a number in this latter range appears to have been about the norm. The practice was for both sides to fire their arrows away at each other, seeking, with their cavalry, to break their opponent's formation and achieve tactical advantage. The battlefield, therefore, was largely static and it would have been a perilous enterprise to wheel re-supply transport into the arena. The role of the archer, for this reason, was largely done once his quiver had been emptied.

The turning point came at the battle of Carrhae in 53 BC, when the Romans, in the person of Marcus Licinius Crassus, were taught a sharp lesson by the Parthians in the use of mobile firepower. It cost the Roman commander his life but it changed little else.

Crassus, a contemporary of Pompeius and Julius Caesar, had been seeking to gain for himself an equal military reputation and, to this end, he precipitated a war against a noble Parthian family, the Suren. He marched into Mesopotamia with, under his command, an unbalanced force of seven legions, about forty-four thousand men, supported by an inadequate number of cavalry commanded by his son, Publius. His objective was Seleucia on the Tigris and, as he crossed the Euphrates to thrust towards it, his troops, unacclimatized to desert warfare and inexperienced, were attacked and encircled at Carrhae by a Parthian force of 1,000 knights, with 10,000 mounted archers. Against all precedent, Sunenas, the enemy leader, had provided himself with unusually large reserves of arrows, carried by 1,000 camels, one camel load for every ten men. On the basis that a camel load weighed some 180 lb, this would have totalled roughly three thousand arrows per camel (at 1 oz per arrow) or 300 arrows per man. The camels offered a great advantage as transport animals since they could keep up, on the march, with the mounted bowmen.

When Crassus realized his danger, he organized his infantry into square formation and his men strove to protect themselves by covering both head and body with their shields against the unending shower of Parthian arrows, which the enemy replenished systematically by withdrawing sub-units from action. We are told that some thirty thousand Roman soldiers were killed by the missiles which rained upon them until nightfall. We can gain a rough idea of the size of the Parthian target, if we judge that each side of the Roman defensive square presented a front of 1,000 yards (45,000 men, arrayed nine deep, one yard frontal per man). It might have been larger but probably not smaller. The Parthians had no need to indulge in rapid fire. If they discharged their arrows at a rate of one per man per minute, releasing 10,000 at a time, they had a sufficient quantity of ammunition to last five hours. We may also assume that each bowman held a further ten to fifteen arrows in his quiver, probably designed in such a manner that they would not tangle on being withdrawn. This arrangement is clearly visible in the longbow quivers recovered from the Tudor ship, *Marie Rose*, and displayed in the museum at Portsmouth dockyard.

A tactic of the mounted archers of the Middle East, traditionally attributed to the Parthians but more probably initiated by the nomads of the steppes, was to pretend flight while firing back over the hindquarters of their mounts. This came to be known as a 'Parthian' shot, or, more commonly today, 'a parting shot'. Again, this was probably equally as much a device to escape unscathed, having exhausted their supply of arrows, as to entrap the enemy. One Roman general, Ventidius, is credited by Frontinus with being particularly successful in finding an answer to this swift, long-range attack.[28] He used to allow them to come within 500 paces and then, 'by a rapid advance, he came so near to them that, meeting them at close quarters, he escaped their arrows, which they shot from a distance'.

The defeat of Crassus at Carrhae dealt Roman prestige in the Middle East a stunning blow but, strangely, the technique which brought the victory and which introduced a mobile long-range weapon, with plentiful, almost unlimited, reserves of ammunition, was not yet adopted. By the time of the Jewish War of AD 67, however, when the trend had veered towards siege warfare, the importance attached to firepower is clearly recognizable in the forces deployed. The reinforcements which Titus brought from Greece to swell the army of his father, Vespasian, brought the total Roman expeditionary army to three legions, the Vth, Xth and XVth; 18 cohorts of infantry, totalling 18,000 men; and supporting levies amounting to roughly 5,000 cavalry and 11,000 archers, provided by local kings, Antiochus, Agrippa, Soaemes and Malchus of Arabia.[29] The number of bowmen in this instance, which included mounted men as well as foot soldiers, suggests that their value by now had been truly appreciated.

The role of the archer was multifarious. Julius Caesar, who seemingly used them but little in Gaul, employed them more frequently in North Africa, in one instance mingled with slingers to provide a protective screen against cavalry attacking his flanks; in another, spaced out among his mass of heavy infantry 'at definite points throughout the line, but chiefly on the wings'.[30] On another occasion he posted them with slingers in the course of a cavalry attack,[31] and again, in Gaul, while on the move he used them either with the vanguard of a column or as its flank guard.[32] Vespasian (AD 69–79) handled them more freely and in larger numbers. At Jotapata, he positioned his Arab bowmen and Syrian slingers, with his mini-catapults and supported by artillery, to withstand the Jewish onslaughts on his ranks.[33] At

Tarichaeae, he detached his archers to seize a flanking hill feature from which to give covering fire for his assault on the town.[34] During the siege of Jerusalem Titus deployed them effectively in street fighting, 'placing his bowmen at the ends of the streets and taking his own stand where the enemy were thickest'.[35]

Before completing this glance at the employment of the sling, the sling-staff and the bow in the role of supporting weapons, there is one other addendum which needs to be considered, namely, the rarely mentioned *amentum*. Ammianus Marcellinus refers to it obliquely when describing a battle with the Goths in AD 378. 'The Romans held the upper hand,' he tells us, 'since no bullet from the thong of a slinger, or any other missile when hurled, missed its mark.' The word *amentum* is usually taken to mean the thong by which a javelin, spear or arrow was given a whirling flight and propelled on its way by hand, with proportionately added momentum. The range is not stated but, in trained hands, we may guess it to have been as much as 200 ft.[36]

Distances covered by the range of small-arms weaponry employed by light and mounted infantry in Roman times may broadly be categorized as illustrated in Table 4. Inevitably, in their search for something that could reach out and disrupt, at a distance, the mass of enemy infantry before it reached close contact with them, soldiers demanded increasingly better performance from their weapons, both in range and in the nature of the projectiles they discharged. Originally, wrote Heron, the construction of these engines was developed from handbows. As men were compelled to project by this means a somewhat larger missile and one of greater range, they increased the size of the bows themselves and of their springs.[37]

TABLE 4

	Weapon	Distance (in yards)	Reference
a.	Sling	30–40	See fn 10, p. 157, but may be greater
b.	*Amentum*	65–70	A sling-arrow: see above
c.	Sling-staff	200	Vegetius, II, 15
d.	Longbow	230–50	Payne-Gallwey, *op. cit.*, Part VII, 20
e.	Comp. bow	360–400	*ibid.*

In this manner, just as the crossbow was born from the composite bow, the *gastraphetes* rapidly evolved, developed from the concept of the crossbow. The *gastraphetes* may be regarded as the first example of non-torsion artillery. In essence, the machine comprised a variety of parts which enabled a man to operate a bow of such power that, without them, he would have been utterly incapable of doing so. In the main, it comprised three elements. First, the bow, which provided the propulsive mechanism; second, the stock, to which the handbow was firmly

Auxiliary archers providing supporting fire in battle

fitted; and, third, a sliding part which could be moved freely, backwards and forwards on the stock, in a dovetailed groove. This part was also equipped with a shallow channel to accept the missile and a trigger mechanism to release it. The rear end of the stock was shaped concavely so as to permit an archer to press his body down upon it when drawing his bow, and thus derive added power. Vegetius provides clear evidence that the Romans employed the crossbow in their armoury. Crossbow archers, he explains, not entirely convincingly, for we have already seen light supporting weapons deployed in a variety of other ways, were normally sited in battle at the rear of the infantry:

> This was also the post of the archers who had helmets, cuirasses, swords, bows and arrows; of the slingers who threw stones with the common sling or with the *fustibalus*; and of the *tragularii* who annoyed the enemy with arrows from *manuballistae* [handbows] or *arcuballistae* [crossbows].[38]

In due course, the size and strength of the crossbow grew beyond the ability of man to operate it in the normal manner. It was then provided with a stock as a firm base, and with a winch mechanism to prepare the weapon for firing. The practice was for the archer to place the nose of the stock on the ground and, while holding the machine in a vertical position, to turn the winch and draw back the bowstring.

This machine provided the prototype of the original non-torsion *ballista*, which, in its turn, was developed into the more powerful torsion-motivated weapon. Marsden estimates that some *gastraphetes* were designed with a bow length of as much as 15 feet, capable of firing a shot of 40 lb in weight a distance of between 200 and 300 yards.[39]

Again, there was a quest for improved performance, the drive being provided by Greek engineers. They appreciated that, if the size of the missile were to be increased and the force of propulsion enhanced, the arms of the bow on the existing machine did not possess sufficient power to generate the energy required. The means by which they decided on the principle of torsion-propelled artillery remains obscure. They may perhaps have been driven to it from a consideration of the part played by animal sinew in the manufacture of a composite bow. However it happened, the spring of the bow was now replaced by two torsion operated lateral arms, which applied a much greater tension to the bowstring while maintaining, in most respects, the working shape of the non-torsion catapult. This, in essence, was the design of the Roman *ballista* and its counterpart, the cart-borne *carroballista*, the mechanical operation of which it is not proposed to discuss here in detail.[40]

Ballistae were manufactured in different sizes for use in differing military situations, the smallest, as we have seen, being not much larger than a heavy

A ballista *in action, depicted by the Ermin Street Guard. In this modern simulation, torsion is applied to the throwing arms by coiled rope. In ancient times it is likely that animal gut or horse-hair were used. On at least one occasion (n. 44), skeins of women's hair were utilized*

crossbow, although its performance in the field was much greater. It possessed arms of about 2 feet in length and was powered variously by skeins, about 4 inches in diameter, of sinew, cord, or horse hair. Indeed, the women of Carthage, in about 148 BC, are said to have sacrificed their hair for the purpose.[41] The largest variety of *ballista* was twice the size of the smallest, with arms between 3 to 4 feet in length, and was powered by skeins of sinew 6 to 8 inches in diameter. Payne-Gallwey experimented with these machines by constructing a small version in his workshops. He found it capable of projecting a stone ball, 1 lb in weight, a distance of 300 to 350 yards. He had little doubt that the large stone-throwing *ballistae* of the Greeks and Romans could cast a round stone, 6 to 8 lb in weight, a distance of 450 to 500 yards. These were often formed of heavy pebbles, enclosed in baked clay and designed to shatter on impact, so that the enemy could not retrieve and make use of them.[42] The logistical implications of this tactic, combined with the need to provide the machines with ammunition of a prescribed weight, size and quality, would have been considerable and would have presented substantial workshop and transport problems.

The production of a repeating catapult by Dionysius of Alexandria in the third century BC is surely worthy of note, if only for its inventiveness. Philon of Byzantium (*c.* 200 BC), who professed to have seen it, comments that its performance was unremarkable and that its range (200 yards) was less than that achieved by its more conventional counterpart.[43] He saw no reason to pursue its development. His main complaint appears to have been that its ability to shoot a large number of missiles rapidly at a given target was a disadvantage, for its drop shot had little spread, either laterally or longitudinally: it was, in other words, too accurate. It may be that he had a reluctance to accept innovation. A similar dissatisfaction was expressed when the Bren light machine gun was introduced during the 1930s as a replacement for the elderly Lewis gun; it subsequently served the British army with great efficiency throughout the Second World War and for many years afterwards.

The existence of a repeating catapult at such an early date gives obvious relevance to Payne-Gallwey's record of an ancient Chinese repeating crossbow,[44] capable of firing ten arrows within 15 seconds, particularly pertinent in view of the recognized flow of Oriental weapons to the Middle East during this period. With such a rate of fire, where 100 men could fill the air with 1,000 arrows in a remarkably short space of time, it was clearly a lethal weapon. Indeed, it was employed to good effect by the Chinese as late as the Sino–Japanese War of 1894–5. The last occasion it was used against the British was at the taking of the Taku Forts in 1860. There is, however, no evidence that this weapon played a part in Roman warfare although the possibility should clearly not be overlooked.

The *onager*, the large siege catapult employed by the Romans, was, in the same manner as the *ballista*, also constructed in varying sizes according to the military requirement. The machine, according to Ammianus Marcellinus,[45] was variously nicknamed *tormentum*, because the tortion provided by the sinew which powered it was produced by extreme twisting; or *onager*, because wild asses, when chased by

Sketch plan of a catapult for slinging stones, its arm being partly wound down

hunters, were said to kick back stones upon their pursuers, 'either crushing their breasts . . . or breaking the bones of their skulls and shattering them'.

Its frame possessed five main components:

a.   a solid wooden base, hewn out of oak, stoutly held together by cross-beams of equal strength, secured by mortise-tenon joints;

b.   a throwing-arm, which Ammianus defined as 'like the pole of a chariot', tapering in width to its top. Here it was fitted either with a throwing sling to hold a stone missile or a spoon-shaped receptacle for larger projectiles;

c.   a skein of elastic material to provide a spring. This was manufactured out of animal sinew, horse-hair or, on occasions, women's hair.[46] Payne-Gallwey records that when the arm of his machine was fully wound, it could not be moved 'even one inch' by three strong men armed with a rope;

d.   a winch mechanism, operated from the rear of the machine and employed to haul the throwing-arm down into a horizontal position. The cable on this device was fitted with a release lanyard;

e.   a heavy reinforced cross bar to stop the forward movement of the arm when it was released. This was padded with 'a great cushion of hair-cloth, stuffed with a fine chaff, and bound on with strong cords' so as to take the impact when the arm was released;

Ammianus defined this process in the following words:

A British (Durotriges) slingstone hoard uncovered at Maiden Castle, Dorset

Then, when there is a battle, a round stone is placed in the sling and four young men on each side turn back the bar with which the ropes are connected and bend the pole almost flat. Then finally the gunner, standing above, strikes out the pole-bolt which holds the fastening of the whole work, with a strong hammer, thereupon the pole is set free and, flying forward with a swift stroke and meeting the soft hair-cloth, hurls the stone, which will crush whatever it hits.

The longer the arm, the greater was the arc it described through the air and, all other things being equal, the further it projected its shot. Added power and increased range were also obtained for the weapon by fitting a sling to the head of the pole, in the same fashion that the performance of a sling was enhanced by being fitted to a sling-staff. When its release was triggered, the arm exploded upright, striking the crossbar with tremendous force and despatching the missile with added velocity. The throwing-arm was thus subjected to great strain and was no ordinary piece of wood: something specially designed was needed but the nature of its precise composition remains obscure. Payne-Gallwey suggests that the arm was manufactured from several spars of wood, glued and held together, with lengths of thick sinew fitted longitudinally, and bound round with broad strips of raw hide. He visualized that, by this means, it would have set as hard as metal.

An *onager* may be expected to have weighed, according to its stone or spear casting capacity, between 2 to 6 tons. It was therefore an unwieldy weapon which presented transportation difficulties. On occasions, it would have been stripped down, packed on to wagons and re-assembled when required for action. Sometimes it was fitted with wheels and directly pulled by teams of oxen or mules. Either way, it was slow-moving and consequently vulnerable. Cestius, commanding a Roman army at Jerusalem in AD 66 at the outbreak of the Jewish War, learnt this lesson to his cost. Finding himself surrounded, with the ground on all sides swarming with Jews, he gave orders for a night withdrawal. All animals were to be slaughtered, 'and even the wagon-horses, with the exception of those conveying the missiles and artillery'. The enemy pressed him hard, however, and compelled him to put on such a turn of speed 'that the soldiers in utter panic dumped the battering-rams and spear-throwers and most of the other engines'. These the Jews promptly seized and were soon operating against their original owners.[47]

The *onager* was a siege weapon and not one generally deployed in fluid operations. It possessed a considerable capability. Josephus[48] relates that the stones it cast 'were of the weight of a talent and were carried two or more *stades*': an *onager* could thus fire a half-cwt stone a distance of 400 to 500 yards. If we accept this statement, and Payne-Gallwey suggests that from the results of his experiments it is not unreasonable, then we must realize that this would not have been its effective range in siege warfare, when employed against high fortifications. A projectile flying through the air describes a parabola, and its range would have been foreshortened by the height of the defences, thus probably bringing the machine within range of enemy small arms.

Josephus's description of their performance at the siege of Jotapata is particularly vivid:

The force of the spear-throwers was such that a single projectile ran through a row of men, and the momentum of the stones hurled by the 'engine' carried away battlements and knocked off corners of towers. There is in fact no body of men that cannot be laid low to the last rank by the impact of these huge stones. The effectiveness of the engine can be gathered by incidents of that night: one of the men standing near Josephus on the rampart got in the line of fire and had his head knocked off by a stone, his skull being flung like a pebble from a sling more than six hundred yards. . . . Even more terrifying than the actual engines and their missiles was the rushing sound and the final crash. There was a constant thudding of dead bodies as they were thrown one after another from the rampart. Within the town rose the terrible shrieks of the women. The whole strip of ground that encircled the battlefield ran with blood, and it was posssible to climb up the heap of corpses on the battlements. The din was made the more terrifying by the echoes from the mountains around. . . .

There is continuing debate about the manner in which these weapons were deployed. According to Vegetius,[49] there were ten *onagri* to each legion, a scale of one per cohort, and one *carroballista* per century, but it is militarily difficult to believe that they could have been usefully handled individually in this manner; the distribution provided by Vegetius may perhaps simply represent a legionary scale of entitlement. The principles of concentration of force and of firepower would by themselves have demanded that weapons of this category be 'brigaded', either by cohorts or, particularly in the case of *onagri*, by being grouped as legionary artillery. There are very few phases of war in which they could have been employed which would not have required some such organization.

There are numerous examples of this truth. Caesar, defending a hill against Belgic forces in 57 BC, dug himself in and concentrated his artillery 'at both ends of each trench, to prevent the enemy from using their numerical superiority to envelop his men from the flanks' .[50] In AD 14, Germanicus marshalled his artillery to cover his crossing of the Adrana river.[51] At the siege of Jotapata, mentioned above, Vespasian assembled 160 'engines for throwing stones and darts round about the city'. On the line of march, artillery, as positioned by Vespasian in the Jewish War and by Arrian in his campaign against the Alani in AD 134, moved *en bloc* behind the auxiliary cavalry and infantry cohorts, but in front of the legionary soldiers[52] and readily under the hand of the commanding general. Again, in the set-piece battle, as we have seen above, it was frequently mustered with the reserves, to the rear of the mass of infantry, or deployed to protect the flanks. Thus, in each phase of Roman war, in column of march, in defence, in set-piece battle, in siege warfare and in the assault, artillery weapons were grouped for operations, an entirely logical arrangement, since it is not easy to judge that a solitary *carroballista* could have offered much benefit to a century.

This leads us, therefore, to the question of the command structure and the administration of the artillery in earlier times. Even in the case of one legion, where ten *onagri* and sixty *carroballistae* were involved, the numbers concerned were considerable. In an army where two, three, four or even more legions were involved, the total of artillery would have been huge. There must, for this reason, have been some chain of command, some *principalis*, the equivalent of a commander, Royal Artillery, through whom the army commander could have passed his orders. The equally large administrative problems of attached artificers and workshops responsible for the maintenance of these weapons in the field would have been closely linked with this requirement. Again, the evidence for such a *principalis* is slight,[53] but it is not easy to believe that the commanding general could efficiently have passed his orders downwards to legions and cohorts in any other manner.

Finally, the numbers of men required to operate these weapons is very relevant. A *carroballista* would have required a junior commander, a gun-aimer, who probably did not have responsibility for loading the missile, one or two men on either flank of the weapon to turn the winch, one or two animal handlers, and perhaps two more men to re-supply ammunition, assuming that reserves of missiles were taken on to the field of battle, making a total of ten. It would also have been important for each artillery piece to have had some senior soldier responsible for its maintenance and knowledgeable of its frailties.

Vegetius[54] has suggested that a team of eleven men was required to operate and maintain each weapon and, although this figure is sometimes questioned, he may not be far wrong. In the case of an *onager*, particularly a large one firing heavy missiles, an even larger gun team would have been needed. The total number of artillerymen per legion, without taking artificers into account, would thus have been in the neighbourhood of 650, roughly of cohort strength. Clearly, many of these men would have been specialists and must have existed as such before the second and third centuries AD. It is strange, therefore, that, even in those late years, there are still only tenuous references to their existence (the *ballistarii*). The *Notitia Dignitatum* records a parade state of Tarrutenius Paternus (*c.* AD 150) in which the gunners are included among the *immunes* – men excused fatigues, guard duties and the like.

Thus, to sum up the true state of the disposition of the artillery weapons. Despite the clear statement by Vegetius that one *carroballista* was allocated to each century and one *onager* to each cohort, it would have been practical, for command and administrative reasons, for them to have been concentrated at unit and formation level, namely, the light weapons at cohort level and the heavy artillery at legionary headquarters. Operationally, everything points to the artillery having been handled in this manner. Apart from the military sense of the idea, it would greatly have assisted the training and allocation of specialist artillerymen and would have economized on the need for artificers and the use of workshops. It is notable that after *c.* AD 300, all recognized artillerymen (*ballistarii*) were designated to special artillery units. There is, however, no evidence to suggest that this is the way it happened at any earlier date, despite the logic of the arrangement.

CHAPTER SIX

Waterborne Operations

The war was pushed forward by land and sea; and infantry, cavalry and marines, often meeting in the same camp, would mess and make merry together.

Tacitus, *The Agricola*

'. . . and had not certain persons in their envy of [Caesar] . . . forced him to return here before the proper time, he would certainly have subdued all Britain together with the other islands which surround and all Germany to the Arctic Ocean, so that we should have had as our boundaries for the future, not land or people, but the air and the outer sea.'

Mark Antony in Dio's *Roman History*

The author of the above words about Caesar, Cassius Dio,[1] was a historian whose *Roman History* covered the years from the foundation of the city by Romulus up to his own times, the end of the third century AD. He thus wrote basking in the glow of his knowledge of imperial achievement, which enabled him to put a vision of the future into the mouth of Mark Antony, when recreating the latter's supposed speech to the Forum upon the assassination of Julius Caesar in 44 BC. Dio's phrasing hints at an early Roman ambition to extend their nation's influence to the very edges of a flat-earth world, where the land mass terminated, suspended in mid-air, and an endless ocean reached to the edges of a distant horizon. Whether or not Caesar nurtured such ideas, they certainly would not have been considered strange by Trajan, who visited the Persian Gulf in AD 116. After making a slow passage down the Tigris, and, 'when he had learnt its nature [the Persian Gulf] and had seen a ship sailing to India',[2] Trajan remarked that he would certainly have crossed over too, if he were still young; but he was destined never to see Rome again and he died in the following year.

Under Hadrian (AD 117–38), who presided over the heyday of imperial Rome, the northern frontier of the Roman world ran broadly along the axes of the Rhine and Danube rivers, with the exception that, east of Pannonia, for good defensive reasons, it deviated abruptly northwards to incorporate the Dacian salient, before resuming its course to the northern shores of the Black Sea, the Pontus Euxius. At the eastern extremity of the latter, the line now turned southwards, to pick a path through the fertile valley of the Euphrates river and reach a point north of Sura. Here, it turned again to head directly for the Red Sea, south of Eilat, in the

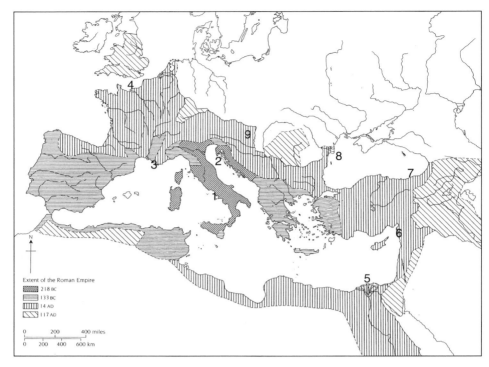

1. Misenum (senior)   4. Boulogne   7. Classis Pontica
2. Ravenna   5. Classis Syriac   8. Classis Moesica
3. Forum Julii (Fréjus) 6. Classis Alexandrina 9. Classis Pannonica

The growth of the Roman empire, showing principal naval ports (from The Oxford History of the Classical World*)*

Gulf of Aqabar. In broad terms, the Middle Eastern frontiers of the Roman Empire at this time encompassed the areas of modern-day Turkey, Syria, Israel and Jordan.[3] From the Gulf of Aqabar, the frontier proceeded westwards to include Egypt and the prosperous agricultural belt of North Africa's Mediterranean coastline, until its course was halted by the Atlantic Ocean. The western boundary of the empire now ran northwards, along Europe's western seaboard, to meet once more with the Rhine delta, embracing in its progress a large swathe of ancient Britain.

At the heart of the Roman Empire lay the Mediterranean Sea, the arrogantly named *Mare Nostrum*, Our Sea, which provided access, through the Pillars of Hercules,[4] to the inhospitable waters of the oceans which lay beyond. This area, together with the southern flank of Rome's African possessions and client states, where a sea of arid desert reached distantly towards an uneven skyline, provided to a considerable degree the trouble-free frontiers of 'air and ocean' pictured by Dio. Elsewhere, as we have seen, they were firmly delineated by the Rhine, the Danube, the shores of the Black Sea and, to a minor degree and at varying times, by the lines of the Euphrates and the Tigris.

In the late empire Rome eked out its slim military resources by using vexillations of up to 2,000 men to supplement her frontier garrisons. This papyrus fragment from Egypt bears the names of Diocletian and Maximian in 'monumental' letters, and 'Vexillation of Leg. V. Macedonica'. It provides a record of a mobilized detachment of this Danubian legion in Egypt, intended for a public notice or for carving an inscription on stone

Rome could neither have won nor held her great empire had she not possessed a navy, trained and of sufficient size both to ensure her domination of the Mediterranean and Black Seas and to enable her to succour and communicate with her widely deployed armies in the field; although, in truth, the numbers of these were not great. 'Almost every nation under the sun bows down before the might of Rome', Josephus quotes Agrippa as saying *c*. AD 40, in a speech aimed at turning the Jews from the folly of war with such a formidable enemy;[5] and then, speaking particularly of North Africa and Asia Minor, he continued,

> none of them [the nations] could resist Roman skill at arms. This third of the whole world, whose peoples could hardly be counted, bounded by the Atlantic

and the Pillars of Hercules, and supporting the millions of Ethiopia as far as the Indian Ocean, is subdued in its entirety . . . and unlike you they take no offence when given orders, though only a single legion is quartered in their midst.

Reading these words, it is tempting to reflect upon the equally moderate forces deployed by Great Britain in the occupation of her empire, in particular the Indian sub-continent, whose population of 385 million people, comprising numerous castes and tribes, was peacefully governed by a highly skilled civil service, supported by 52,000 British regular soldiers, a further 135,000 troops of the Indian army, with an additional 45,000 men contributed by the Princely State Forces. The paramount responsibility of the military, apart from occasional duties in aid of the civil power, rested with the control of the North-West Frontier Province and the defence of that strategically vulnerable area in the event of exterior aggression. There was also a police force of modest strength, totalling 203,000 officers and men. Much of the governance of the sub-continent remained in the hands of Princely State rulers (client states, it might be argued), carefully and discreetly guided by British residents. As in the case of Rome, there was a major factor which rendered possible this 'light hand on the tiller' approach to government, namely, the benefits of *Pax Britannica* enjoyed by the governed and the sense of security they provided. There were, as well, other important influences, but they are not relevant to this chapter.

Parallel with the expansion of their empire, the British government, like the principate, directed its attention to the control of an increasing number of key strategic areas so that speedy reinforcement and the free flow of resources, along carefully planned and prepared lines of communication, might be guaranteed whenever an emergency demanded.

In the same manner, Rome's naval control of the Mediterranean formed a vital part of her strategy, although it must at once be said that, until the third century AD, her fleets were rarely confronted by opposition on the high seas. The execution of this policy rested in the hands of two senior squadrons, *classes praetoriae*, based individually at Ravenna and Misenum and charged with the task of guarding, respectively, the eastern and western seaboards of Italy. Of these two naval bases, Ravenna, built on stilts in what Strabo called 'the marshes', and lying in the north-eastern corner of Italy overlooking the northern extremity of the Adriatic Sea, was a large and busy city.[6] It was linked by canal to the mouth of the Primaro and to Spina. The harbour, frequented by the Roman navy and initiated by the emperor Augustus (31 BC to AD 14), was sited about 3 miles outside the town. It quickly emerged as an important centre, with detached flotillas at Aquiela, Ancona and Brindisi in Italy, another at Salonae (Split) in Dalmatia (Yugoslavia) and a further two in Greek waters.[7]

Misenum, a deep-water harbour established *c.* 22 BC, quickly became the foremost naval station in the empire. It stood at the end of the northern cape of the Bay of Naples, where its proximity to the emperor at Rome ensured its importance. The hinterland adjoining Misenum, according to Strabo, contained 'the most blest of all plains'; and the bay itself 'was . . . garnished by residences and plantations, which, since they intervene in unbroken succession, present the appearance of a single city'.[8] Many of these properties, with their fruitful, fertile lands, were the possessions of past and present imperial families.

In the fullness of time, a detachment of the Misene fleet was stationed at Ostia, at the mouth of the Tiber, with duties which appear to have included the transportation of the imperial family, officers and staff, the receipt of despatches and the task of carrying the emperor's orders to his commanders overseas. Indeed, it was from Ostia that Claudius, accompanied by his staff, an element of his Praetorian Guard and members of his family, had sailed to be present at the final conquest of Britain in AD 43. He descended the River Tiber by boat and, having twice been forced by storms to take shelter, then followed the coastline to Massilia (Marseilles) from where, 'advancing partly by land and partly along the rivers, he came to the ocean and crossed over to Britain'.[9]

The Misene fleet possessed subsidiary naval bases in Sicily, Sardinia and Corsica, together with another at Forum Julii (Frejus), in Narbonesis. Of these, the latter was probably the most important, by virtue of its valuable tactical location adjoining the mouth of the River Rhône and lying close to Marseilles, an important, well-fortified harbour, the possession of the people of a client state, the Massiliotes.[10] Tacitus records how Valerius Paulinus, in the first century AD, having seized the naval base at Forum Julii at a time of national crisis, despatched a flotilla of fast galleys 'to the Stoechades Islands which belonged to Marseilles'.[11] It is probable that a detachment from Frejus was also located at Arles, a Ligurian township situated at the point where the Rhône divides to form its estuary, north-west of Marseilles, with the task of establishing a military presence and ensuring the peaceful passage of river traffic to and from the north.

Ostia was an important port, particularly for the reception and storage of grain imported from overseas and destined for Rome. Strabo, nevertheless, describes it as harbourless. Silt carried down to the sea by the many mountain streams which feed the Tiber had, even then, created a bar across the estuary, forbidding access to large merchant ships. These were thus denied the opportunity of making their way upstream to the emporium at Rome, which lay 22 miles distant, and were compelled either to lighten their cargoes by partly off-loading into warehouses at Ostia, or to exchange cargoes off-shore, while anchored in what often proved to be troublesome tidal waters. Nor did rivercraft find the passage to Rome particularly easy. Laden barges plowing their way against the sometimes powerful waters of the Tiber almost invariably needed to be towed from the river bank by teams of buffalo; in other instances, horses were employed.[12] This factor should be noted, for it provides an answer to many of the military supply problems we have been considering and was common behaviour in ancient times, when men, rather than animals, were employed as the towing medium, and is still to be seen in many parts of the world today. Boats with a burden of 6 tons, provided with a strong towing team, could make 20 km a day upstream; downstream, as much as 30–35 km was possible.[13] The practice was further complemented by the fitting of inflated bladders to rivercraft to facilitate their movement through shallow waterways.[14]

The apparent lack of attention paid by early writers to the logistical problems of their times, particularly those affecting military operations, has often been a matter of frustrated remark by historians glancing at this period. Many of the

River scene with coracle and fishermen riding inflated pigskins. Relief from the Palace of Sennacherib (705–681 BC), now in the British Museum

ways and means of the ancients, such as those employed for the accumulation of reserves of foodstuffs, the evacuation of wounded (of whom there must have been many in the bloody and numerous campaigns of the era) and the forward provision of grain, remain obscure. It may be that the reason for this is to be found in the fact that the techniques employed were so glaringly obvious and so fully established in the public mind that they were deemed to be unremarkable. Strabo,[15] in a paragraph describing the river system of Gaul (about which more than average data are available, mainly due to the detail of Caesar's *de Bello Gallico*) makes it clear that he, at least, appreciated their potential in this regard. Some of the rivers flow down from the Alps, he wrote,

the others from the Cevennes mountains and the Pyrenees; and some of them are discharged into the ocean [*the English Channel*], the others into Our Sea. Further, the districts through which they flow are plains, for the most part, and hilly lands with navigable watercourses. The river beds are by nature so well situated with reference to one another that there is transportation from either sea into the other; for the cargoes are transported only a short distance by land, with an easy transit through plains, but most of the way they are carried on the rivers – on some into the interior, on others to the sea. The Rhône offers an advantage in this connection, for not only is it a stream of many tributaries . . .

but it also connects with Our Sea, which is better than the outer sea and traverses a country which is the most favoured of all in that part of the world.

Strabo[16] also describes how traffic was portaged for onward transhipment, either down the Garonne and the Dordogne (which share the same estuary), the broad waters of the River Loire or the River Seine, from whence it began 'its journey down to the ocean and to the Lexobi and the Caleti; and from these people it is less than a day's run to Britain'. Let us consider the direction and potential of these waterways in rather more detail.

The River Rhône rises from a glacier in south central Switzerland and, after entering Lake Geneva, continues on a zig-zag course through the Jura Mountains until it meets with its tributary, the Saône, at the city of Lyon. It then flows directly into the Mediterranean Sea and is the only significant river in Europe to do so. At suitable places along its course – and this was a feature which applied to most riverways of this scale in Gallic-Romano times – land and waterway routes interconnected to form transhipment points at which cargoes switched from river to land, or vice versa, to undertake the next stage of their journeys. Seyssel, for example, described by Chevallier as 'the outer harbour of Geneva on the open Rhône flowing down to the sea',[17] provided a vital nodal point of this nature and conveys a strong impression of the weight of traffic which then plied these routeways, penetrating to the heart of Europe. Seyssel lay at a point where two main lines of communication converged: one coming from the north from Germany and the Rhine, the other descending from the Great St Bernard and passing along the left banks of Lake Geneva and the Rhône. From it, it should be noted, for it puts its use in perspective, stone was later to be shipped downriver for use in the building of Lyon; to it, in another age, came amphorae from Italy and from Iberia.

The Saône, a tributary of the Rhône, is itself a substantial river which flows some 300 miles before melding with the waters of the main stream at Lyon. It rises near Epinal in south-west Lorraine, a watershed area which it shares with the Moselle, and, meandering southwards, passes through Chalons-sur-Saône, Tournus, Macon and Villefranche before joining with the River Doubs at Verdun-sur-le-Doubs, north of Lyons. It appears historically to have been a busy trading river, with many craft plying its waters. It will be recalled (Chapter 3 pp. 53–4) that the Aedui, in 60 BC, had lost a war with the Sequani over the payment of tolls on the Saône and had called upon Rome for military assistance to re-establish their 'rights'. The cause of the disagreement appears to have been rooted in a convoluted quarrel arising from Aeduan customs duties levied on exports of 'the finest of salted hog-meat', for which their neighbours had found an agreeable market in Italy.[18]

Traffic reaching the headwaters of the Saône was, in ancient times, confronted by a short overland portage between the Saône and the Moselle, ultimately connecting with the Rhine, whose waters in turn, flowing westwards through Coblenz, Cologne and Arnhem, provided access to the North Sea. A paragraph by Tacitus records how, during the reign of the emperor Nero (AD 54–68), Vetus had intended to link these two riverways but had been frustrated by local jealousies.[19] To keep the troops busy, wrote Tacitus, the imperial governor of Lower Germany, Pompeius Paulinus, finished the dam for controlling the Rhine, built 63 years previously by Nero Drusus. His colleague in Upper Germany, Lucius Antisus Vetus, planned to build a Saône-Moselle canal. Goods arriving from the Mediterranean up the Rhône and the Saône would thus pass via the Moselle into the Rhine, and so to the North Sea. Such a waterway, joining the western Mediterranean to the northern sea-board, would eliminate the difficulties of land transport. But the imperial governor of Gallia Belgica, Aelius Gracilis, jealously prevented [him] from bringing his army into the province he governed. . . .

Today, the vision of Lucius Antisus Vetus has surely been vindicated, for a thriving canal now exists between Metz and Coblenz, permitting passage of Rhine barges as large as 1,500 tons in capacity between the two rivers .

Nero Drusus, mentioned above by Tacitus, died in 9 BC.[20] During his service in Germania he constructed a canal which connected the Waal, a tributary of the Rhine, with the Ems, thus enabling rivercraft to avoid the hazards of the North Sea. Some sixty years later, his purpose was carried yet further forward by Corbulo who, for a similar reason of keeping 'the troops occupied', constructed a 23 mile long canal between the Meuse and the River Rhine. He was granted an honorary Triumph by Claudius in recognition of this work and others of the same nature, for he was renowned for his enterprise. Tacitus records that his exhausted troops secretly appealed to the emperor, begging that he should allocate honorary Triumphs to his generals *before* giving them command and thus restrain their energies!

The River Doubs, a tributary of the Saône, which in its upper reaches meanders through gorges and is partly torrent, cannot be omitted from this pattern of riverways. It follows a course of 267 miles, after rising near Mouthe in the Jura Mountains. Although today it has been canalized, in order to bypass otherwise difficult falls and rapids, it would not have been easily navigable in Roman times beyond a comparatively short distance from its mouth. Besançon, the largest town of the Sequani, may have fallen within this range, as might Belfort, an ancient town which occupies the Burgundy Gate and dominates the strategic passage way connecting the Rhine valley with the Paris basin and, consequently, the River Seine. It was at Besançon, where the Doubs forms an almost complete circle around the town, that Julius Caesar established his first winter base in 58 BC, after his brief and successful campaign against the Helvetii.

The number of vessels on this great network of riverine routeways, and available to the military for charter or requisition, would obviously have been considerable. For confirmation, we need look no further than an event in 52 BC, when Labienus[21] was despatched by Caesar, together with four legions, to seize Lutetia, a settlement of the Parisii situated on an island in the Seine. His arrival had been anticipated and, upon reaching the edge of the water opposite his objective, he discovered that all craft and other means which might have assisted his passage across had sensibly been removed. He therefore commenced to throw a causeway across the marsh separating him from the township, laying down a foundation of fascines (bundles of long sticks) and other materials, but the idea

Scene from Trajan's Column showing Roman infantry crossing a pontoon bridge. Each soldier carries his pack and other kit on a stick, in the fashion decreed by Marius

proved unworkable. Then, abandoning the project, he marched upriver to Metlosedum, the town of a neighbouring tribe, also standing on an island in the Seine, where he found an assembly of fifty boats. Floating these downstream, he lashed them together and formed a bridge which enabled him to get troops across the river and into his objective without further ado.

It is logistically significant that, throughout his campaigning in Gaul, Julius Caesar never strayed far from a network of a major river with its tributaries. His operations in 53 BC, for example, which ranged between Rheims, the capital of the Remi, Autun, the capital of the Aedui, and Lutetia, were plainly based on the facilities offered by the River Seine, with its major tributaries. Of these, he employed two for his purpose: the Yonne, which approaches Paris from the south, and the Marne, which, with its parent river, the Seine, rises on the Plateau de Langres, north of Dijon, and then sweeps in a great northerly curve to pass through Rheims, before joining the main stream, a few miles east of Paris. In the following year, Caesar based his activities upon the Loire and its tributary the Yeure, while still maintaining contact with his main grain providers, the Aedui. Their fertile territory in the Autun district was ideally located in a position of great influence, lying as it did at the heart of the French pattern of riverways. During this

A side view of Caesar's Bridge across the Rhine as depicted in the early eighteenth century by William Stukeley in his Itinerarium Curiosum

campaigning season, Caesar established a rear headquarters at Nevers, itself on the river, and carried his operations to Bourges and Orleans, before returning to Alise St Reine, 30 miles north-west of Dijon, to confront Vercingetorix at Alesia.

In due course, and with their customary thoroughness, the Romans provided a pattern of roadways to complement their river systems. One such road[22] which is significant, for it hints at a master plan which refuses to be intimidated by water obstacles of whatever size, was initiated by Claudius during his years as emperor. It extended from Chartres to the coast of the Cotentin peninsula where, on the other side of the Channel, it lay opposite the start of Vespasian's tactical road from Hamworthy to Bath, or, alternatively, the Isle of Wight and the estuary shared by the Rivers Stour and Avon. Of these two rivers Sean McGrail[23] has commented that their tributaries are wide-ranging and that, 'by their use and with a portage from the Wiltshire Wylye to the Somerset Frome, there is theoretical access to the Bristol Channel and the Mendip and Cotswold regions'. It is noteworthy that, as late as 1623, it was being visualized that 'the Avon might be made as navigable as the Thames' from Salisbury to the sea.[24] The justification for the proposition was that wood had to be carted to the city from distances of 18 to 20 miles, while 'a boat would carry as much as 20 wagons and 80 horses and such boats would bring up sea coal and send downstream corn, beer, bricks, stones, tiles and charcoal. . .'. Both these Roman options, it should be noted, pointed to Bath as their ultimate destination, linking with the Bristol Channel and the River Severn. Bath, situated on the Bristol Avon, could have been a key military administrative base and was the possible location of a military convalescent centre.[25]

If we return to Chartres, and extend Claudius's Roman road in exactly the opposite direction, we discover it runs to Lutetia, the settlement seized by Labienus a century earlier and the origin of modern-day Paris. Even at that moment in history, this was a commanding centre, with Roman roads radiating outwards to

Senlis, Meaux, Melun, Dreux and Chartres. Across the Seine, its fingers reached to an area bounded by the Meuse and the Escaut, the strategic importance of which, as a corridor to and from the Rhineland, has been periodically re-discovered over the centuries by successive generations of soldiers. This avenue led to the Roman garrison on the Upper Rhine, deployed on the empire's northern frontiers and covering the Danube Gap, the land bridge lying between the sources of the Rhine and the Danube. In support of these operations, particularly on the Lower Rhine and the river's tributary, the Ems, there existed the *classis Germanica*, a flotilla whose formation was dated by Tacitus to 6 BC.[26]

The Roman fortifications designed to close the Danube Gap were not set in place until the Antonine era, when a palisade was erected, covered by defensive positions much the same as those laid out by Hadrian along the Wall in northern Britain; but it was Augustus (31 BC to AD 14) who preoccupied himself for nearly twenty-five years in consolidating the Pannonian frontier (Northern Yugoslavia to the head of the Danube valley) and in pushing forward the Roman frontier to the Danube along the whole length of the river. It is likely that two Roman flotillas on the Danube, the *classes Pannonica* and *Moesica*, had their origins in his reign. Equally, it is probable that both the IXth Legion *Hispania*, moved from Pannonia for the invasion of Britain in AD 43, and Aulus Plautius, the regional governor of the province, were selected for this duty because of the specialist knowledge they had accumulated during riverine operations on the Danube in previous years. The line of the IXth Legion's advance, first along the line of the Thames to the Medway,[27] and then northwards up the east coast of Britain, while almost certainly working closely with the *classis Britannica*, would seem to bear this out. The ancient Roman canal linking the Wash, via the Witham, with the Trent at Torksey, may also be dated to this time, the middle of the first century AD.

Downstream from Pannonia, the responsibilities of the *classis Moesica*, apart from the security of the Danube delta, are deemed to have included patrolling the western shores of the Euxine (Black Sea), with the influence of the fleet extending as far south as Byzantium (Istanbul), and penetrating the Straits of Salonika.

During the reign of Vespasian (AD 69–79), the province of the Danube became increasingly a source of strength for Rome's eastern legions, providing both men and material for her army on the Armenian front. The supply line initially ran from the Danube estuary across the water to a port at Trapezus, on the south-eastern shore of the Black Sea. This was also the destination of supplies of grain, hides and fish provided for the same purpose by her ally, the client kingdom of the Bosphorus. Goods from here were transported by the *classis Pontica*, which also was charged with the task of subduing the many pirates who, at that time, were savagely exploiting merchants trading in those waters. Supplies landed at Trapezus were then conveyed over a difficult mountain road to Roman forces on the Euphrates and, at a later date, the Tigris. Trapezus, therefore, served a vital need, both as a port and as a naval base which, according to Josephus, in the first century accommodated a fleet of forty ships.[28]

By the end of the second century, the Pontus ports on the southern shores of the Black Sea had taken on a less critical military role, their logistical mantle having been assumed by the Levant, which, as a consequence, had been correspondingly enhanced in importance, as had the roles of the *classes Syriaca* and *Alexandrina*, jointly responsible for the naval security of the eastern Mediterranean coastline. This operational shift was achieved by the garrisoning of the Euphrates frontier in Cappadocia by Rome and by the extensive development of a system of military highways into Asia Minor. The expanding military district, of which Antioch had become the administrative centre, was now served by a fast-developing port at Seleucia, situated at the mouth of the Orontes, on the Mediterranean coast, north-east of Cyprus. Antioch also possessed the added logistical advantage of being historically linked to Palestine and Egypt by routes of high antiquity. The affect of this swing in the military balance was that the *classis Pontica*, while seemingly still maintaining a toe-hold in Trapezus, appears to have made greater use of Cyzicus, in the Straits of Salonika. The latter port, according to Strabo, at one time boasted 200 ship sheds and harbour facilities to match.[29]

It has already been remarked that the Roman navy held unchallenged seapower throughout its years of empire. There was no enemy of comparable size, nor with equivalent resources, to oppose it. It thus had small need either to train or equip for specialized set-piece battles on the high seas. Its purpose, rather, was to concentrate its efforts on fundamentally routine tasks such as anti-piracy control, and the provision of communications and transport facilities, at varying levels, for the imperial armies it served. Vessels used by the imperial navy included the *liburnian*, a fast, two-banked galley which took its design from local craft operated by the piratically minded inhabitants of the Dalmatian coast and offshore islands; and such other variations as the trireme, quadrireme, quinquereme, together with one 'six', the latter being shown as based at Misenum as the flagship of this, the senior squadron.[30] It was also, doubtless, heavily used by the imperial family.

A trireme was constructed with three banks of oars, but it does not follow that the larger categories increased in the same manner. Indeed, it is generally assumed that they did not do so and that a 'four', for example, had two banks of oars with two men per oar and a 'six' either three banks with two men per oar or, more likely, two banks with three men per oar. The numbers of these various categories within the navy is obscure but an archaeological study of ships' names on Roman navy graves at Misenum and Ravenna has provided the data for Table 5.[31] Thus, proportionately, trireme names discovered at both Misenum (48 per cent) and Ravenna (66 per cent) predominate, conveying the impression that this was the favoured vessel for both 'home fleets'. The size of the Misenum total, compared with that of its east-coast partner, underlines the high status awarded to the western Mediterranean squadron. The *liburnian* category, on the other hand, although not recorded here in any great number, appears to have been the workhorse of the provincial fleets. The *classis Alexandrina*, as one example, has been attested with nothing else.[32]

The emperor Augustus, frequently styled as the founder of the Roman navy, concentrated on the military virtues of speed and manoevrability when designing his ships. The need for these essential qualities had been impressed upon him in 31 BC, when he defeated Mark Antony in a naval battle at Actium. In the interests of both operational and administrative efficiency, he favoured a standardized fleet, limited both in numbers and in size differential. Vessels with these same qualities

TABLE 5

Archaeological study of ships' names on Roman Navy graves at Misenum and Ravenna

Grading	Misenum	Ravenna	Total
'Six'	1	–	1
'Fives'	1	2	3
'Fours'	9	7	16
'Threes'	52	22	74
Liburnians	13	2	15
Combined Total	76	33	109

were brought to Caesar's aid by Decimus Brutus in 56 BC, and turned an uncertain situation into a decisive victory over the Veneti on the Atlantic coast.

The Venetians, seemingly at the end of the campaigning season according to Caesar,[33] had seized some Roman soldiers sent out for grain and afterwards detained two envoys despatched by Crassus to demand their release. The purpose of the Gauls was to gain the release of their own hostages held by the Romans. When the news reached the Roman commander he was well on his way to Italy; he tells us that he at once foresaw the need for further warships and issued instructions for these to be constructed on the Loire (probably on its upper reaches, at Roanne, west of Lyon) and for the local recruitment of captains and seamen. They were then sailed down the river, to the estuary. These ships, however, proved inadequate for the task and he decided he must wait for his fleet 'to be assembled and brought up'. Directly they hove in sight, some two hundred and twenty enemy ships sailed out of harbour to confront them and neither the commander of the fleet, Brutus, 'nor the military tribunes and centurions in charge of the individual ships could decide what to do or what tactics to adopt'. Finally, they attacked and brought down the enemy's sails with grappling poles; after that, Caesar grudgingly remarks, 'it was a soldier's battle in which the Romans proved superior, especially since it was fought under the eyes of Caesar and the whole army. . .'.

Dio, however, tells a different story. Caesar, he relates,

constructed in the interior the kind of boats which he heard were of advantage for the tides of the ocean, and conveyed them down the river Liger [Loire], but in doing so used up almost the entire summer to no purpose. For their cities, established in strong positions, were inaccessible and the oceans surging around practically all of them rendered an infantry attack out of the question, and a naval attack equally so in the midst of the ebb and flow of the tide. Consequently Caesar was in despair until Decimus Brutus came to him with swift ships from the Mediterranean. . . . For these boats had been built rather light in the interests of speed, after the manner of our naval construction, whereas those of the barbarians surpassed them very greatly both in size and construction.[34]

The Venetian ships were fitted with both sail and oars, and were designed to be sufficiently heavy to compete with the high seas prevailing on the Atlantic coast. They could maintain their manoeuvrability and mobility providing there was a wind. When this fell, however, their weight was such that they could no longer 'be propelled as they had been with the oars and, because their of great bulk, stopped motionless'. Brutus chose this moment to attack, in a manner graphically described by Dio, which underscores the wisdom of Augustus's choice of design two or three decades later:

> . . . falling upon them, he caused them many serious injuries with impunity, delivering both broadside and rear attacks, now ramming one of them, now backing water, in whatever way and as often as he liked, sometimes with many vessels against one and again with equal numbers opposed, occasionally even approaching safely with a few against many. . . . If he found himself inferior anywhere, he very easily retired, so that the advantage rested with him in any case. For the barbarians did not use archery and had not provided themselves beforehand with stones, not expecting to have need of them. . . .

The Roman navy, both before and after its reconstruction by Augustus, largely functioned as a supporting arm of the army. Every crew was treated as a century of the Roman army, commanded by a centurion, the rank and file including not only the oarsmen, but also marines, archers, catapult operators and other specialists. Indeed, Casson[35] judges that the naval hierarchy was probably charged with nothing more than the operation and maintenance of the ship, while the army took care of everything else, including shore administration and responsibility for the fighting component.

This method of behaviour was particularly evident in riverine operations, where a fine line existed between the requisitioned rivercraft employed on the Rhône by Caesar for multifarious military purposes and the *lusoriae*, the small light galleys which, by the fourth century AD, had been adopted as the standard unit for use on such major rivers as the Rhine and the Danube. Neither would have held more than a handful of men for their own defence and, when operating in hostile territory, would have needed land-based protection. An account of Julian's operation on the Euphrates in AD 363, narrated by Ammianus Marcellinus,[36] clearly recognizes this fact when he tells us that 'the fleet, although the river along which it went winds with many a bend, was not permitted to lag behind or get ahead' of the army. Again, when Julian is compelled by circumstances to discuss the abandonment and destruction of the fleet,[37] he stressed the advantage 'that nearly 20,000 soldiers would not be employed in transporting and guiding the ships, as had been the case since the beginning of the campaign'. This appears an abnormally large number of men to devote to such a purpose but emphasizes the point that the need was recognized. It is possible to see these rivercraft as broadly the equivalent of those provided to the British Army by RASC/RCT Water Transport Companies in recent wars.

In Roman amphibious operations, the direct influence of the army in naval affairs is immediately visible. A classic example again involves the ubiquitous Caesar. In order to gather intelligence before setting forth on the invasion of Britain in 55 BC, he despatched a warship *under command of a soldier, Gaius Volusenus*, to make a general reconnaissance and then hasten back with his report. Volusenus, a military tribune, was Caesar's cavalry commander in later campaigns. In this, his first invasion across water, Caesar made three fundamental mistakes:

Firstly, he set fourth in two echelons, the infantry from Boulogne and the cavalry from Ambleteuse, six miles further north. Any sailor could have told him of the dangers of such a plan. The cavalry transports set sail too late and were carried back to the shore by the strength of the tide.

Secondly, he did not sail until about 25th August, proposing to stay three weeks. He therefore risked being trapped in Britain by the autumn equinoctial gales, a hazard of which Claudius, a century later, was seemingly well aware. It was, said Caesar, a matter about which Romans knew nothing, apparently ignoring the fact that Roman merchants had been trading across the Channel for many years and must have been aware of these dangers, as indeed should his ally and cavalry commander, Commius of the Atrebates.

Thirdly, he aimed to live off the land upon arrival but his enemies were equally determined that he should not be allowed to do so. This error was compounded by his inexperience of the vagaries of Channel weather for when several days of continuous storms blew up, wrecking many of his transports, his 'whole army was thrown into great consternation' and his administration was near collapse.

But to dwell on these failings unduly (from which, as ever, he extricated himself with skill) would be to deny the brilliant planning of the assault landing.

Caesar assembled for his 'reconnaissance' eighty transports, which he considered sufficient to convey two legions, together with an unstated number of 'warships, which he assigned to the *quaestor*, the generals and the officers of the auxiliary troops'. The latter vague description presumably also included a complement of auxiliary troops, such as archers and slingers, to operate with the mounted contingent when it arrived. Additionally, he possessed several scouting vessels[38] and eighteen further transports which, as we have seen, were marshalled separately, some miles away along the coast, and allotted to the cavalry. When Caesar returned to Britain in the following year, he came with an increased total of five legions and two thousand cavalry, transported by more than eight hundred ships. It is thus not unreasonable, on this first landing, to judge that with two legions, supporting troops and weapons, he brought with him some two hundred and fifty vessels.

He set sail at midnight. It was conventional practice, and had been for many years, that every ship should carry lights at night to identify itself when travelling in company: flagships carried lights on the stern.[39] As early as 204 BC Scipio had ruled, for his expedition against Carthage, that 'warships were to carry one light each, transports two, and the flagship would be marked at night by carrying three'. It was customary for the commander-in-chief to lead the way. In 48 BC, while campaigning in North Africa, Caesar set forth by sea 'instructing the rest of the captains to steer . . . by the lantern of his ship'.[40] Each sailing ship and warship towed astern at least one ship's boat, and these were to prove invaluable assets during the assault landing. Landfall was made at about nine o'clock but, as may be imagined, the fleet straggled far behind their leader and Caesar was compelled to ride at anchor until three o'clock in the afternoon to await their arrival. He took advantage of this enforced interval to assemble

> the generals and military tribunes and, telling them what he had learned from Volusenus, explained his plans. He warned them that the exigencies of warfare, and particularly of naval operations, in which things moved rapidly and the situation is constantly changing, required the instant execution of every order. On dismissing the officers he found that both wind and tidal current were in his favour. He therefore gave the signal for weighing anchor and, after proceeding about seven miles, ran his ships aground on an evenly sloping beach, free from obstacles.[41]

The above quotation is important, not so much for what is said but for what is left unsaid. Caesar was rightly and obviously using his flagship as a command ship for the operation but his account of these events in *de Bello Gallico* does not make it clear whether, when he assembled his 'generals and military tribunes', these officers were already members of his shipboard party or had been summoned from nearby vessels. The first alternative is more probably correct. One matter is, however, absolutely certain: they could not have been travelling with their regiments and formations, for these were yet to take six hours to assemble. Thus, it appears that his commanding generals and senior staff were sensibly moving alongside him, in the modern manner, and using the same 'orders group' procedure later practised by Vespasian when advancing in column of march.

A further piece of knowledge we are denied by Caesar is the manner in which his scattered task force was re-assembled after arrival, parallel with the shore and positioned in echelons as they would have needed to be, ready for the assault on the beaches. In this instance, with a mere 250 ships participating, the problem would not have been of the same magnitude as in the following year when eight hundred vessels and more were involved, nor as in the Claudian invasion of AD 43, when some thousand took part. It would, however, have been very necessary, after the problems encountered during the approach of his task force to its objective, for the commander-in-chief to re-form his regiments into their assault formations before going ashore. Moreover, this essential task could only have been achieved with speed and efficiency if the whole manoeuvre had been pre-planned. Vessels and groupings of vessels would have needed to carry identifying marks, probably coloured pennants, and carefully rehearsed signal procedures would have had to be in position.

A great variety of signalling methods was available. Diodorus[42] records how, in 307 BC, Demetrius used a burnished shield as a heliograph to order the fleet into action and the message 'was flashed by each ship to the next'. The Athenians in 410 BC hoisted a purple flag for going into action,[43] thus hinting that other flags, or other hues and dimensions, might have conveyed a wide variety of meanings. Dain[44] confirms this latter point and reveals that for general signalling a type of semaphore was used, for which the instructions were to

> signal with the flag by holding it upright, dipping it from right to left and then shifting it back again to right or left, waving it, raising it, lowering it, removing it from sight, changing it, switching it around by orienting thé head now in one direction and now another, or using flags of different shapes and colours as was the practice amongst the ancients.

It was also commonplace for ships to fly flags to indicate their country of origin, or at least to identify the fleet to which they belonged. At the siege of Marseilles in 49 BC, Caesar relates[45] how the flagship of Decimus Brutus was identified and attacked by a couple of enemy triremes because it was 'conspicuous by its ensign'. Many Roman coins feature shipping flying different shapes and sizes of flag. An issue made at Side displays a small rectangular flag attached to a horizontal bar. Roman imperial coins issued by Trajan and Severus show a similar flag flown by what is taken to be the imperial flagship.[46] It might be the commander-in-chief's personal banner. Several ships are illustrated displaying two or three such banners, but one or two of these could be related to the formations or units being transported in them. Regimental and sub-unit standards, probably displayed upon the poop deck on such occasions, would have played little part in these naval activities but would have come into prominence during the landing operation, as beach 'assembly points', for example. It is to this phase of the assault that we now turn.

When Caesar made landfall off the coast of Britain, he found the 'enemy' already awaiting his arrival, posted on the cliff tops and ready to deluge the shore with javelins and other missiles. They had seemingly been alerted to their danger by tribal intelligence received from the continent. When he ultimately moved off after the arrival of his main body, to seek a more secure landing place, his opponents despatched their cavalry and chariots to anticipate his movements. Their infantry followed close behind, while maintaining visual contact with his fleet. By this means, as the first Roman ships approached the shoreline, the Britons were already in position to give them a rough welcome. Caesar's account of what happened next fully describes the grave difficulties which confronted him:

> The size of the Roman ships made it impossible to run them aground except in fairly deep water; and soldiers, unfamiliar with the ground, with their hands full, and weighed down by the heavy burden of their arms, had at the same time to jump down from the ships, get a footing in the waves and fight the enemy, who, standing on dry land or advancing only a short way into the water, fought

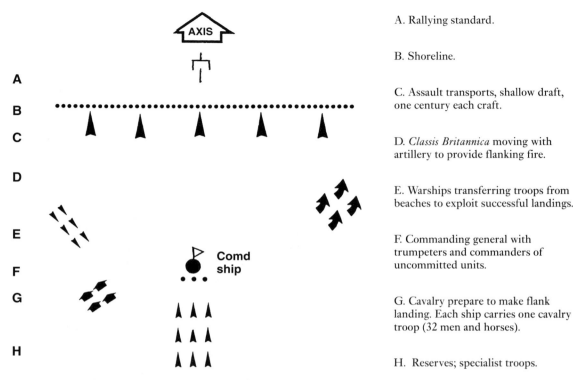

A. Rallying standard.

B. Shoreline.

C. Assault transports, shallow draft, one century each craft.

D. *Classis Britannica* moving with artillery to provide flanking fire.

E. Warships transferring troops from beaches to exploit successful landings.

F. Commanding general with trumpeters and commanders of uncommitted units.

G. Cavalry prepare to make flank landing. Each ship carries one cavalry troop (32 men and horses).

H. Reserves; specialist troops.

Waterborne assault landing (source: Caesar, de Bello Gallico, *iv, 23–6)*

with all their limbs unencumbered and on perfectly familiar ground, boldly hurling their javelins and galloping their horses, which were trained for this kind of work. . . . These perils frightened our soldiers, who were quite unaccustomed to battles of this kind, with the result that they did not show the same alacrity and enthusiasm as they did in battles on dry land.[47]

Caesar, from his command ship anchored offshore and, we may imagine, surrounded by a cluster of vessels which, apart from his staff, would have included reserve troops, auxiliaries, supporting arms, light galleys and ships' boats (the latter being a general purpose boat, also acting the part of signals 'galloper'), was concerned by this lack of movement. He was seeking a quick success. The sun, on 25 August, would have set at about seven o'clock and last light would have occurred about an hour later. Time was not a factor to be ignored. He reacted at once, ordering his warships to move to the flank of the assault beach and run ashore, taking the auxiliaries with them. From this stable position they were now enabled to direct fire from slingshot, archers and artillery on to the exposed flank of his British enemy, who, rendered uncertain by this unexpected new danger, paused momentarily to consider what they should do about it.

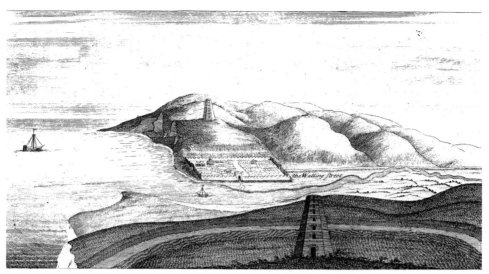

William Stukeley's imaginative view of Roman Dubris *(Dover), drawn in the early eighteenth century*

There now occurred a much quoted incident: the standard-bearer of the Xth Legion, one of Caesar's favoured formations, took advantage of this break in the fighting and, 'after praying to the gods that his action might bring good luck to the legion, cried in a loud voice: "Jump down, comrades, unless you want to surrender our eagle to the enemy. I, at any rate, mean to do my duty to my country and my general." ' His comrades, inspired by his bravery, at once poured off the ship to join him on the beach, and their example quickly spread to neighbouring landing craft. But this was not the end of the difficulties of the Roman task force.

Much of the great success of the Roman soldier on the battlefield lay in his knowledge of close formation fighting: the concept of the battle drill involved, the training of the soldiers, the interlocking of their shields, the nature of their equipment and the design of their short stabbing broadswords, all had their part to play in the technique. As the men now struggled ashore out of the water, wet and isolated, bereft of any formation, the Britons, with their traditional long, slashing swords, inflicted heavy casualties among them: but we are not told how many. The legionaries, pressed hard by the enemy, did not find it easy, in the general mêlée, to get a foothold on the beach, nor keep their ranks. They fought their way to any standard they could find and assembled around it and, ultimately, some semblance of order began to emerge. Caesar, in his command ship, was watching these events carefully. He now gave orders for all light galleys and ships' boats to be loaded with reserve troops and he deployed these shrewdly, with the purpose of exploiting areas of success where he could see them and bolstering any sector which might appear in danger of defeat.

In due course, as the legions fought their way out of the water and formed their battle lines, their heavy infantry gained enough space to mount a charge. The Britons withdrew in the face of this onslaught and, as they retreated, Caesar was

The Roman pharos *(lighthouse) at Dover*

driven to regret the absence of the cavalry which had failed to arrive, for he had no means of pursuing his enemy and sealing his victory.

There are questions to be answered, arising from Caesar's account of this expedition. The first concerns the use of warships to provide artillery fire and the scale of issue of weapons to each vessel. It would have been sensible, if it were planned to use them in this manner, for gunships to be prepared, equipped with twenty or thirty weapons, so that they might deliver concentrated fire. This would have economized on the numbers of vessels employed and, militarily, would have been more effective. We can only conjecture whether or not it happened in this way. Caesar's account of his North African campaign against Labienus in 46 BC certainly conveys the impression that warships carried more than one of these *carroballistae*, when he relates that, in order to improve his defences at Ruspina, 'he brought missiles and artillery from the ships into the camp'.[48] Warships, in later years certainly, and possibly even at this early date, carried many means of destruction. During the reign of the emperor Constantine, a particular type of galley, known as the *dromon*, was fitted with siphon-like flamethrowers, and could frequently boast more than one to a ship. These weapons proved to be devastatingly lethal. Each comprised a copper-lined tube of wood, fitted on its inboard end to an air pump, into which a substance known as 'Greek fire' was poured. When the pump was activated, the contraption squirted forth a stream of fire over its target. It was also common practice for naval vessels to carry a catapult as armament. This was employed to throw baskets of red-hot coals or jars of flaming pitch at an enemy craft.

Secondly, we are left to wonder what happened to Caesar's badly wounded, of whom there must have been quite a number. We may hazard a guess that ambulance ships formed part of his command group and that, while walking wounded would have been treated ashore, within their regiments, severe cases would have been repatriated to Boulogne with all speed. A *medicus* (ship's doctor) formed part of the crew of most warships of the imperial Roman navy.

Finally, in his detail of the landing in 55 BC, Caesar makes no mention of transport animals, except to say that 'he had come without most of his heavy baggage'. This implies that he did bring *some* heavy baggage and that, therefore, he must have had the means of carrying it, together with any grain he might have been able to forage, once ashore. It would thus seem that Caesar's task force contained a Q Echelon of supply ships. These administrative vessels may well have formed part of his command group and, during the assault phase, would probably have been anchored to the rear, but out of range of the busy command ship, with its reserves and supporting arms and vessels.

Model of a Roman grain ship

Seaborne landings of this nature, in the face of organized opposition, were rarely undertaken by the Roman army and it is this fact which makes the detail of Caesar's planning so remarkable, especially when compared with modern practice. It refutes the claim of the military historian, Major General J.F.C. Fuller, that the Roman general's 'two invasions of Britain were amateurish in the extreme'.[49] There were, of course, administrative failures in the conduct of his operation, partly due to over-optimism, partly because of excessive faith in the influence of his Celtic friend, Commius, the Atrebatan. But the successes and failures of Caesar's two invasions of Britain must surely have laid down military guidelines for future expeditions: there can be little doubt they would have been carefully noted and followed by the Claudian army a century or so later.

In earlier chapters we have already noted Fuller's view – somewhat harsh we have suggested[50] – of aspects of Caesar's command ability, provoked largely by his seeming lack of a wagon train and an organized commissariat. Fuller felt this latter omission to be the most serious of the Roman's deficiencies because it led to undue reliance upon his nation's allies 'to provide the vehicles needed to supply' him. But it would appear from this comment that he did not appreciate the presence of the waterborne option. Indeed, in the light of Caesar's skilful employment of the riverine routeways of Gaul, with his rear headquarters more often than not located in river-based tribal capitals, it is not easy to agree that such a commissariat did not exist. Clearly, he was not the originator of the concept of river supply, it dated to ancient times, but in Gaul he appears to have adopted and adapted it to good effect and, in doing so, to have been fortunate (or was it not entirely fortuitous?) that his chief ally and provider of grain, the Aedui, occupied the main watershed area of his targeted conquest.

In later years, the practice of using riverine military supply routes was developed to an extraordinary extent by the Romans. Ammianus Marcellinus[51] has recorded one occasion which occurred in AD 363, when Julianus Augustus was campaigning in Mesopotamia along the line of the Euphrates:

> While [the emperor] was giving audience his fleet arrived, equal to that of the mighty King Xerxes, under the command of the tribune Constantianus and Count Lucillianus; and the broad Euphrates was almost too narrow for it, consisting as it did of a thousand cargo-carriers of varied construction, and bringing an abundance of supplies, weapons and also siege-engines; there were beside fifty warships and an equal number which were needed for making bridges.

Marcellinus's detailed account of this occasion emphasizes a scale of operational planning of considerable ability.

CHAPTER SEVEN

Siege Warfare

For here lofty embankments were being raised, there others were filling up the deep ditches; elsewhere long passages were being constructed in the bowels of the earth, and those in charge of the artillery were setting up their hurling engines, soon to break out with deadly roar.

Ammianus Marcellinus xxiv, 4

The stronghold is a *machan* overlooking a kid tied up to entice the Japanese tiger.

Major General Orde Wingate[1]

A siege, in its most simple terms, is the investment of a fortified place by an encamped force with the purpose of either seizing it or compelling its surrender. There are thus three elements to a siege, of which the first is a fortified place, which may be a town, castle, port, hillfort or some other such feature from which a siege might be repelled, its location being carefully selected with an eye to water supply, ground, communications, tactical importance and defensibility, dependent upon the nature of the operation. In addition, there are the besieged and the besieger, these two components being mentioned individually because, although their presence is obvious, their motivation and purpose is necessarily different. A siege generally arises from the spontaneous action of those under attack, who may see it either as a means of warding off a stronger, better prepared enemy or, alternatively, as an integrant part of a premeditated system of offensive defence.

A classic example of this latter tactic was the system of burghal hidage[2] originated and adopted by Alfred the Great (AD 849–99) to frustrate the ravages of Viking raiders in southern England. His scheme was based on the creation of thirty *burhs*, each chosen for its prime tactical position and each situated within 20 miles of the other, thus enabling the local population, in time of war, to withdraw within its walls, leaving behind no stocks of food and destroying any standing crops which might bring comfort to the enemy. A specific number of hides (a taxable area of land often about 120 acres in size) were assigned to each *burh*, the occupants of each hide being required to provide one man to fight upon the ramparts in the event of an attack, the total defence force being calculated on the basis of one man to every five yards of rampart. A further man, contributed by each hide, was additionally required to march with the *fyrd*,[3] whenever it might be called out. Each *burh* stood ready to move to the support of its neighbour should the need arise.

A more modern example of offensive defence of this nature is to be found in the 'stronghold' concept devised during the 1941–5 Burma campaign by Major-General Orde Wingate, commander of the famous Chindit long-range penetration columns. He saw the 'stronghold', a firmly held and fortified administrative base, as a '*machan* overlooking a kid tied up to entice the Japanese tiger'. Its ideal situation was 'the centre of a circle of 30 miles radius consisting of closely wooded and very broken country, only passable to pack transport owing to great natural obstacles'. Ideally, it should have a 'neighbouring friendly village or two and an inexhaustible and uncontaminatable supply of water'. Wingate visualized his strongholds creating such situations that his enemy would inevitably be drawn towards them, into battlefields of his own choosing, in difficult terrain, where their forces would be worn down and their lines of communication open to harassment by his mobile fighting columns.

It is possible to get a glimpse of some such tactic in the defensive operation conducted in 52 BC by the brave but unfortunate Gallic leader, Vercingetorix, prior to the siege and capture of Avaricum (Bourges) by Julius Caesar and again at Alesia.[4] The vision fades, however, almost as soon as it emerges for, despite setting in place arrangements which might have made his victory possible, the young Gaul failed to carry them to fruition.

Vercingetorix was heading a rebellion against Roman rule. Caesar's *de Bello Gallico* does not provide us with a detail of the strength of the army his young opponent had gathered around him, beyond stating that he had won support for his cause in districts as far apart as Sens, Paris, Poitiers, Cahors, Tours, Evreux, Limoges, Angers and the whole Atlantic seaboard and that, armed with this power, he had ordered each of the allied states to furnish a specified number of troops without delay, together with a fixed quota of weapons, paying particular attention to cavalry needs.[5] Clearly, the numbers thus accumulated were considerable. In his campaign of 58 BC, Caesar estimated the strength of the Helvetian army then confronting him as numbering 92,000.[6] It is unlikely that Vercingetorix's army numbered less than this, on an estimated basis of 10,000 men donated by each of the tribes marching with him. On the other hand, a scratch force of this nature would not have possessed the disciplined military skills of the Roman legions opposing it, a fact of which both commanders would have been well aware.

Vercingetorix opened his campaign by summoning tribal leaders to a council of war, to gain their approval of his operational plan, which combined a scorched-earth policy with a design for defence. He outlined to them its background, which was directed towards denying any form of sustenance to the Roman legions. Supplies were to be centrally stored in defended areas where they would not fall into enemy hands. Fields were to be cleared of grain and fodder; nothing was to be left standing. All villages and farms along Caesar's line of march, wherever his foragers might conceivably reach, were to be burnt to the ground. 'When our lives are at stake,' he instructed his followers, 'we must be prepared to sacrifice our private possessions.' As an additional measure, the 'scorched earth' thus created was to be heavily patrolled by cavalry, of which he possessed large numbers and the Romans comparatively few, to ensure that not a single enemy foraging party, if any did emerge from their columns, might return alive to base.

Next, Vercingetorix turned his attention to the physical defence of his area. All towns, 'except those rendered impregnable by natural and artificial defences', were to be burnt. If this were not done, he warned,

> they may serve as refuges for shirkers among our own numbers, and give the enemy the chance of looting the stores of provisions and other property they contain. You may think these measures harsh and cruel, but you must admit it would be a still harsher fate to have your wives and children carried off into slavery and be killed yourselves, which is what will happen if you are conquered.[7]

These drastic plans were unanimously approved by the tribal elders, with one exception. The Bituriges had not appreciated that their town of Avaricum, one of the finest in Gaul and the heart of their tribal defensive system, was destined to be destroyed. They now pleaded for it to be spared. It could easily be held, they said, because of its natural strength, 'for it was almost completely surrounded by an area of river and marsh, in which there was only one narrow opening'. Vercingetorix was at length prevailed upon to grant their plea, although he was at first totally opposed to doing so. He committed 10,000 infantry to its defence. The response of the Bituriges was total. Next day, the number of fires burning on the skyline revealed to Caesar that more than twenty of their towns had been set alight. The remaining cities, spared destruction, prepared themselves for the forthcoming Roman assault.

Thus far, so good; but beyond this point the Vercingetorix plan fell into disarray. This, almost certainly, was brought about by the speed of Caesar's movement. Vellaunodunum, a town of the Senones in central France, lying between Sens and Orleans, succumbed to his attack within three days. He had initially intended to bypass it but was wary of leaving such a fortification across his line of communication. He then moved with equal rapidity against Cenabum, a town of the neighbouring Carnutes, lying on the site of modern Orleans, and arrived outside its walls two days later, while the tribal leaders were still collecting troops for its defence. Caesar plundered and burnt the unhappy town without compunction and then crossed the Loire, into the territory of Bituriges, to lay siege to Noviodunum, some 18 miles north-west of Bourges. Its terrified citizens yielded to him without resistance and he at once turned to the task of subduing Avaricum.

The crossing of the Loire by Caesar, and his march on Avaricum, drew Vercingetorix towards him but the latter carefully avoided physical engagement. He followed the Romans 'by easy stages' and then encamped deep in a marshy forest area, 15 miles from the town, presumably hoping to entice his enemy into this unfamiliar ground. If this were so, he was unsuccessful, for Caesar opened his assault upon the city without delay.

A logical and popular tactic, frequently contemplated early in such an operation, might have been for Caesar to isolate the water supply to Avaricum. It was a ploy used by Alexander the Great at the siege of Babylon (143–142 BC), when he is said not only to have diverted the course of the Euphrates from the city but also to have penetrated its defences by following the line of the dried up river bed he had thus created.[8] Caesar used a similar tactic in 51 BC, during his war with another Gallic tribe, the Cadurci, when besieging Uxellodunum. In the instance of Avaricum, standing as it did in marshy country at the confluence of the Yevre and Auron rivers, it was well provided with a secure water supply and Caesar, if he had contemplated action along these lines, would have been disappointed. Instead, he concentrated his attack upon the city fortifications, employing classic siege procedure.

The walls of Avaricum were constructed in the Gallic style, with layers of stone alternated with heavy balks of timber, these being laid in parallel lines, mortised together, with the gaps thus created being filled with rubble. Caesar described the technique as well adapted for the defence of a town: 'The masonry protects it from fire, the timber from destruction by the battering-ram, which cannot either pierce or knock to pieces a structure braced internally by beams running generally to a length of 40 feet in one piece.[9] The whole circuit of the wall was topped by towers, furnished with fighting platforms and protected externally by dampened hides against attack by fire.

Caesar was a master of the art of siege warfare, which normally followed a recognized pattern of events. The first and obvious phase was to impose a blockade, with the purpose of starving the garrison into submission. The second phase provided a natural corollary to this; a line of entrenchments, known as a circumvallation, was dug around the objective, out of range of bow, sling or catapult shot, with the dual purpose of denying access to the enemy fortification and of providing to the army encamped outside its walls shelter from surprise attack from within. These encompassing entrenchments, seemingly found unnecessary by Caesar at Avaricum, frequently covered considerable distances: 9 kilometres in the case of Aemilianus Scipio at Numantia in Spain,[10] and as much as 17 kilometres by Caesar at Durazzo, during his Illyrian campaign against Pompey.[11] The besieged at that time constructed '24 hill forts on a 14 mile circuit, within which they could safely go out to forage'.[12] The third phase of a siege comprised the development of a further line of entrenchments (a contravallation) which faced away from the place under attack and protected the rear of the besieging army, with its stores and workshops, from possible attack from outside. Simultaneously preparations for the assault were pressed forward, their purpose and magnitude being dependent upon the scale of the obstacles to be overcome.

Because of the natural defences surrounding Avaricum, and perhaps because of a lack of Gallic firepower, the Romans appear to have had no outer ditch and rampart obstacles to overcome. In other circumstances, these might have been found. Caesar's first task, therefore, was to entrench himself in a location from where he might cover the narrow gap which provided access across the marshes to the city. He then began to build a ramp, some 330 feet wide, which he angled towards its battlements. At the same time, he set in position a line of mantlets to provide cover for the soldiery engaged in this construction work and erected two tall defensive towers from where covering fire might be delivered upon any enemy opposition directed at his working-parties, in particular the enemy superstructure of defensive turrets which lined the entire length of their wall.

His opponents countered every move he made. As his ramp approached and grew higher, providing his assault towers with greater height, ultimately reaching

A cavalryman running down a barbarian, from the first-century tombstone of Flavinus, signifer *in the* ala Petriana *from Corbridge, England*

80 feet, they responded by extending upwards the fire-platforms constructed within their wall turrets and frequently made sorties by day and night to set fire to his workings. When the Romans threw ropes, with scaling hooks, up on to the walls, the enemy made them fast to windlasses and wound them up, sometimes with their human cargo. When the Romans erected ladders they cast them down. When Caesar constructed underground galleries to enable his men to approach the walls unseen and without danger, they countermined them 'and prevented their continuation by throwing into them stakes sharpened and hardened in a fire, boiling pitch and very heavy stones'. They also undermined his 'terrace'. Caesar noted that the Gauls were expert at tunnelling 'due to experience gained in the extensive network of iron mines to be found in their country'.[13]

While all this was progressing, Vercingetorix, according to Caesar,

> having run out of forage, had moved nearer to Avaricum and had taken command in person of the cavalry and light-armed infantry who regularly fought amongst the cavalry, in order to ambush the place where he expected our men would go the next day to forage.[14]

Caesar quickly took advantage of the enemy leader's absence from his main encampment and slipped away from Avaricum at midnight to conduct a surprise assault upon it at dawn. They had been alerted and he found their army lined up on high ground, almost surrounded by marsh, awaiting his arrival. He wisely determined to withdraw and return his attention to the siege. The enemy, he explains, plainly not wishing to yield them any credit,

> held all the fords and thickets that bordered the marsh, determined, if the Romans tried to force a passage, to overpower them by running down to the attack while they were stuck fast in the mud. Awaiting us at such a short distance, they looked as if they were prepared to fight a battle on more or less equal terms; but their position was so much stronger than ours that this show of courage was clearly a mere pretence.[15]

Vercingetorix also abandoned his mission without success and returned to base to find the tribes angry at his absence at such a crucial moment and complaining that he had chosen a camp-site too close to the enemy for comfort. He managed to persuade them that his choice had been carefully considered and that his action had been in the general interest. Avaricum fell to Caesar a few days later.

What, then, went wrong? Vercingetorix had generated popular support for his rebellion among the tribes. They had provided him with a cavalry wing of considerable size, far in excess of that possessed by the Romans, and a substantial host of tribal infantry, natural irregular soldiers, with a close knowledge of the local terrain. His carefully planned and rigorously executed scorched-earth policy, watched over by intensive cavalry patrolling, had set the stage for success but, almost immediately, it was weakened by the speed of Caesar's advance. The Roman general, in his unceasing quest for surprise, rapidly overran the territory of the Senones and the Carnutes and then crossed the Loire into the land of the Bituriges. He seized, on

Vercingetorix surrenders to Caesar after defeat at the Battle of Alesia

his way, the towns of Vellaunodunum, Cenabum and Noviodunum, even, in the case of Cenabum, as troops were being moved in to provide its garrison.

The capture and subsequent plunder of these places would have helped fill his supply depot with food, at least for some time, thus partially counteracting Vercingetorix's scorched-earth policy, but a chance still remained that it might succeed. The Gaul had wisely avoided close contact with the highly professional Roman army. His task now was to bring Caesar's rolling advance to a halt. His only hope of doing this would have been to put a substantial garrison into Avaricum, bring about a long siege, force Caesar into entrenchments, both circumvallation and contravallation, and then, by dint of vigorous patrolling and local diplomacy, deny him supplies while keeping him fully occupied, both to his front and rear. In short, to bring the Romans themselves under siege. Vercingetorix, by his initial reluctance to put troops into Avaricum, and then only to provide a minimum number, revealed a sad lack of appreciation of these possibilities.

It has already been remarked that events at Avaricum followed the natural course of any place under siege, that is to say, everything depended upon the degree of success of the assault on the walls. Traditionally, this could have been carried out in three ways, or combination of ways: by using heavy, iron-shod battering-rams to breach them, by employing scaling ladders and towers to surmount them and by employing sappers to undermine them.

Ram and tongs; rams were either mounted on wheels or suspended by ropes or chains. They were worked by men protected under sheds and were occasionally equipped with hooks or tongs to seize men or missile-throwing engines, etc.

A battering-ram was either mounted on wheels or suspended by ropes or chains but, either way, it was provided with overhead cover to give protection against missile and blazing pitch attacks directed at it and its handlers by the besieged enemy on the battlements above. Occasionally, the latter brought up a swing-beam, fitted with a grapnel and operated by a counter-weight, which they lowered in an effort to disrupt the operation and seize the siege engines being deployed against them. Livy relates, with a nice touch of humour,[16] how a similar device was used in 214 BC by Archimedes, to help repel a seaborne assault on Syracuse. The grapnel was lowered on to a vessel's bows, the beam of which was then raised sharply out of the water by shifting the counter-weight: the result 'was to stand the ship, so to speak, on her tail, bows in air. Then the whole contraption was suddenly let go and the ship, falling smash as it were from the wall to the water, to the great alarm of the crew, was more or less swamped.'

Some Roman battering-rams are said to have been 100 feet long and to have required 200 men or so to manhandle them.[17] A tree large enough to provide a ram of this length would have probably come from the softwood family and we may judge the beam would have weighed 4–5 tons. On Trajan's Column, Dacians are shown using a ram in this manner against a Roman fort. According to Josephus, the ram was sometimes supported at its point of balance by ropes passing over another balk of timber, itself resting on stout posts fixed into the ground on either side. It was then operated by 'a great number of men' who, with 'a gigantic united heave' swung it repeatedly backwards and forwards, continuously inflicting shocks to the wall with its projecting iron head.[18] One mark of the engine, which later came into general use, took the form of a frame on wheels, in which the ram, suspended from a horizontal beam, was moved up to the walls. The whole contraption was sheltered by a wooden roof covered, as a protection against fire, with clay or damp hides and was nicknamed a 'tortoise' or *testudo*. These machines were not simply deployed at ground level: some varieties, such as those depicted on the Arch of Septimius Severus, were fitted with two or more stories so that the most vulnerable parts of a fortification, high up on a wall, might be attacked.

The task of the ram's head, from which the battering-ram derived its name, was to shatter the wall by the shock of its impact. Another variety of the weapon, known as the *terebra*, the 'borer', and described by Vitruvius, was fitted with a sharp point and was used for making holes in a wall. A further device was a hook attached to the end of a swinging battering-ram: it was used to pull down stones from the top of a wall once it had started to crumble.

Penetrating a wall by means of a battering-ram was necessarily a lengthy operation. The wall of the city of Jerusalem,[19] 15 feet thick and 30 feet high, was 'built of bonded stones 30 feet long and 15 broad, so that it would have been very difficult to undermine with iron tools or shake with engines'. The besieged thus frequently had time for counter-attack and this took two or more popular forms, either undermining the platform upon which the battering-ram was located, or dropping incendiary devices upon it from aloft or by sorties from the sally-port, aimed at its destruction. In the first instance, the technique was to tunnel under the ram, conserving the space thus created with timber props, while filling it with oil-soaked brushwood. The stake-props and brush wood were then set alight, as the miners hastily withdrew to safety. If all went well, the siege engine fell into the blazing hole as the supporting timbers collapsed, consumed by fire.

The ancient Greeks, and following them the Romans, gave considerable thought to the matter of incendiarism as a weapon in this type of operation. Aeneas the Tactician includes much advice in his treatise on *The Defence of Fortified Positions*. If a fire were to be effective, he advises, it has to be inextinguishable. He therefore recommends a formula made of pitch, sulphur, tow and granulated frankincense, the whole to be mixed in sacks of pine sawdust and kept available to be ignited and cast down upon the siege engines of the enemy as the situation demanded:[20] alternatively, he suggests that faggots covered with the lighted mixture should be lowered by rope on to the target below or 'hurled at the approaching engines'. In another paragraph[21] he advises that pestles should be prepared (large weighted darts, some 5 feet in length), covered with the mixture, and kept ready to 'be dropped upon the [enemy] engine as it is being pushed up [and] fashioned so as to stick into it'; and when mining, or counter-mining, dry brush should be retained in the tunnels created, ready to be ignited should the enemy succeed in breaking through.

Flaming missiles, such as arrows, javelins and sometimes shot fired by siege-engines, were frequently employed in the assault by the besieging troops. Fire extinguishing arrangements and other means of countering the fire were, therefore, important:

If the enemy tries to set anything on fire with a powerful incendiary equipment you must put out the fire with vinegar, for then it cannot easily be ignited again, or rather it should be smeared beforehand with birdlime (*a sticky material smeared on trees to trap birds*), for this does not catch fire. Those who put out the fire from places above it must have a protection for the face, so that they will be less annoyed when the flame darts toward them. . . .

If there were any wooden towers, or if a part of the wall is of wood, covers of felt or raw hide must be provided to protect the parapet so that they cannot be ignited by the enemy. If the gate is set on fire you must bring up wood and throw it on to make as large a fire as possible, until a trench can be dug inside and a counter-defence be quickly built. . . .[22]

Sorties by those under siege, directed at setting fire to the engines of war, were no less dramatic. Josephus has left us with a graphic description of one such event, led by three Jewish soldiers, which took place during the siege of Jerusalem in AD 70:

They dashed out as if towards friends, not massed enemies; they neither hesitated nor shrank back, but charged through the centre of the foe and set the artillery on fire. Pelted with missiles and thrust at with swords on every side, they refused to withdraw from their perilous situation until the engines were ablaze. . . . The Romans tugged at the battering-rams while the wicker covers blazed; the Jews, surrounded with the flames, pulled the other way and, seizing the red-hot iron, would not leave go of the rams. From these the fire spread to the platforms, outstripping the defenders. Meantime, the Romans were enveloped in flames and, despairing of saving their handiwork, began to withdraw to their camps.[23]

Whatever tactic was selected as the means of assault upon a fortification, it called for considerable courage from the participants and generally resulted in a high proportion of casualties. Livy,[24] writing of an incident during Scipio's siege of New Carthage in 210 BC, has equally vividly portrayed the dangers which confronted escalating parties armed with ladders. In the defence of the city, he wrote,

neither men nor missiles were as effective as the walls themselves, for few ladders were long enough to reach the top, and the longer the ladders the less secure they were. The first man up would find himself unable to get over, others would be mounting behind him and the ladder would break under their weight. Sometimes the ladders stood the strain, but the height made the climbers giddy and they fell. When everywhere ladders were breaking and men falling, and success was bringing added keenness and courage to the enemy, the recall was sounded, and thus the besieged were given the hope of an immediate respite from trouble and strife. . . . But scarcely had the din and confusion of the first assault died down when Scipio ordered fresh troops to take over the ladders from their exhausted or wounded comrades and make yet another, more vigorous attempt. . . .

A siege. This sketch from Polybius *(ed. 1727 by Chevalier Follard) depicts both defenders and besiegers at work. Note, particularly, the chamber undermining the tower, with props in position and waiting to be set alight*

Sapping was a military tactic already widely practised by the end of the sixth century BC. The besieged employed it as a means of counter-attack. For the besieger it had three main purposes, namely, to bring combat engineers, or assault troops preparing to scale the walls, close up to their objective without exposing them to missile attack; to undermine the wall and cause its collapse; and to penetrate under the wall, deep into the heart of the fortification under siege. This latter action frequently stimulated a complex pattern of mining and counter-mining such as occurred under Marcus Fulvius, during the Roman siege of Ambracia, in 189 BC. His assault on the walls of the town had been repulsed by its defenders, who built new walls as fast as he destroyed them. He therefore determined, under cover of mantlets, to tunnel his way through and was making considerable headway until the townspeople became alerted by the steadily growing mound of spoil removed from his hitherto undetected workings. They immediately excavated a trench, inside the wall, across what they judged to be his line of advance. Then, occupying this, they placed their ears to the side of the trench and listened for the sound of digging. When, at length, they judged it was sufficiently close, they burst through upon the enemy miners, assailing them with pick and shovel. They were quickly joined by armed men and the intruders were thrown back to take refuge behind a hastily prepared barricade. A novel device was now employed to drive them from the tunnel:

They pierced a hole in the bottom of a cask for the insertion of a tube of moderate size, and made an iron pipe and an iron lid for the cask, the lid also being perforated in many places. They filled this cask with small feathers and placed it with its mouth towards the tunnel; and through the holes in the lid very long spears, called *sarissae*, jutted out to keep off the enemy. A small spark was introduced among the feathers and they kindled it by blowing with a smith's bellows applied to the end of the pipe. Then, when the whole tunnel was filled with a mass of smoke, and with smoke rendered more pungent by the reason of the foul stench of burning feathers, scarcely anyone could endure to remain inside it.[25]

On occasions, siege operations were highly complex, using every aspect of technique. Polybius records that Philip of Macedonia, at the siege of Echinus in 211 BC, planned a two-pronged attack on the town and constructed two towers for this purpose. In front of each he provided a shelter for the engineers engaged in tunnelling towards the walls and for the protection of those men working the battering-rams with which each tower was equipped. These forward shelters, in turn, were linked rearward with his main camp by roofed underground tunnels,

Shelter providing protection for soldiers during attack or mining operations. Note the spades being used in front

'so that neither those coming from the camp nor those leaving the works should
be wounded by missiles from the town'.[26] They were also interconnected and
accommodated 'three batteries of *ballistae*, of which one threw stones of a talent's
weight [approximately 58 lb] and the others stones of half that weight'.

The two towers, giant contraptions which on other occasions are said to have
been constructed as high as 150 feet, were slowly inched forward as the sappers
beneath them levelled the surface of the ground to make their movement possible.
On the first floor of these, two catapults were located, with 'water jars and other
appliances for putting out fires'. The second floor was manned by archers,
slingers and javelin throwers ready to take on any defenders on the walls who
might intervene to stop the progress of the work. It was also customary to find a
drawbridge at this level, ready to be lowered on to the wall, at an appropriate
moment, when sufficiently close to launch an attack.

The time-scale required to make these preliminary preparations is noteworthy.
Polybius relates that the work carried out, together, presumably, with the
manufacture of the two great siege engines, was

> entirely completed within the course of a few days, as the country around us has
> an abundance of the materials required. For Echinus is situated on the Malian
> Gulf, facing south, opposite the territory of Thonium, and the land is rich in
> every kind of produce, so that nothing was lacking for Philip's purpose.[27]

The administrative, manpower and supply resources needed to complete this task
'in the course of a few days' would have been considerable, but an ability to work
to such timing would have been essential to any commander entering upon a siege
operation. Caesar's adversary, Vercingetorix, for example, would have been wise
never to have lost sight of this fact, for here lay the Achilles heel of any besieging
army. Wingate well expressed this concept when defining the purpose of his
'strongholds'. He saw them as well defended and provisioned sites, suitably
located in relation to his main objective, and as an orbit around which columns of
his brigades circulated, harrying the lines of communication of the Japanese who
surrounded him. Such an outside force, as visualized by him, was essential if the
defenders of a stronghold were to be able successfully to discourage a besieging
army; or, alternatively, the strongholds had to be so well armed, fortified and
provisioned that defeat was an impossibility. The introduction of the artillery
weapon transformed the face of siege warfare.

The first siege in which ancient artillery is known to have participated occurred
in 397 BC. The occasion was the assault on Motya, the Carthaginian island
fortress located at the western end of Sicily, under command of Dionysius I of
Syracuse. The island accommodated both an emporium and a colony. It lay in a
sheltered lagoon and was at that time connected to the mainland by a 1,700 m
causeway which, according to archaeological research in 1962,[28] had seemingly
been breached by the Motyans, presumably in anticipation of the coming assault.
Dionysius appears to have repaired the damage and constructed additional moles,
possibly as platforms for his covering towers and siege-engines, which he used to

A fortified town under siege by catapults and ballistae *(Polybius, ed. 1727)*

drive the defenders off the walls. It was, in essence, an attack from the mainland, across a long reach of water and under cover of light artillery fire, to enable the landing party to get ashore. The tactic thereafter expanded relatively rapidly but its employment was more evident in assault than in defence. The later siege of Halicarnassus by Alexander the Great (334 BC), marked a further development in the employment of artillery as a supporting arm: here, for the first time, the presence of stone-throwing machines was recorded. Alexander, having forced a breach in the wall, came under attack from a sortie from within. He reacted vigorously: large stones were hurled by the engines he had mounted on his covering towers, 'bullets were showered in volleys' and his assailants fled back to the safety of the city, having, according to Arrian,[29] suffered heavy loss.

Ancient artillery weapons were capable of a wide range of performance and, as a consequence, were employed to fulfil a great variety of tasks. In defence, they provided covering fire for sorties by raiding parties and were frequently deployed in counter-battery roles. At Halicarnassus, non-torsion arrow-firers were carefully sited to 'volley straight ahead at the advance guard of the engines' as they were being moved by the enemy into position. Similarly, they were located to cover likely lines of enemy approach to the stronghold they were defending. They were also frequently positioned in casements, on the curtain walls of fortifications, so as to bring down enfilade fire on to the flanks of enemy escalades.

A classic early example of these various defensive techniques was demonstrated by Archimedes, the famous Greek mathematician and inventor, who played an important role in the defence of Syracuse against the Romans in 213 BC. His tactical thinking and carefully constructed fire plan delayed the capture of the city for many months. When at length it fell to Marcus Claudius Marcellus, in the late autumn of the following year, Archimedes was slain by a Roman soldier. He was described by Livy[30] as unrivalled in his knowledge of astronomy but as being even more remarkable as the inventor and constructor of types of artillery and military devices of all kinds, 'by the aid of which, as it were, he was able by one finger to frustrate the most laborious operations of the enemy'.

Archimedes' battle plan was uncomplicated:[31] it called for a keen knowledge of the artillery weapon and was almost entirely dependent upon the delivery of a high volume of concentrated and continuous firepower. The Romans opened their siege of Syracuse by an assault from the sea; they could equally well have conducted a land attack, as they were ultimately to do. The Archimedes plan encompassed both possibilities. As soon as the Roman landing force came into target range at about 400 yards, the estimated maximum range of the Greek heavy artillery, it was his intention, as it advanced, to keep it continuously under fire from the city walls, in the first place using catapults and stone-throwers, shortening his range by adjusting his elevation where necessary and, subsequently, by the use of smaller pieces of artillery. Heavy weapons were necessarily slower to fire, because of the weight of their missiles and the difficulties of handling them: light artillery produced a wider and more effective volume of anti-personnel fire, although it was clearly not so damaging when directed at shipping.

There was, however, a blind spot in these arrangements: the batteries of bolt-shooting and stone-throwing catapults accumulated by Archimedes, mainly sited on defensive fortifications 30 to 40 feet high, were unable sufficiently to depress the angle of their fire to engage the enemy closer than a distance of 60 yards or so. He resolved this difficulty in two ways: firstly and perhaps obviously, although it appeared to cause the Romans some surprise, he arranged for the casemates of the fortification to be pierced with large numbers of loopholes at the height of a man and, at each, he positioned archers with rows of so-called 'scorpions', a small catapult which discharged iron darts. These were sited so as to be able to engage the landing-craft as they arrived, with their scaling-ladders, beneath the curtain wall, while simultaneously enfilading the defensive line of fortifications. Secondly, he set in place on the battlements numerous swing-beam cranes, with stocks of heavy stones (some weighing as much as ten talents) and leaden ingots, to be dropped on to ships and troops below, as they gathered for the assault.

The Archimedes fire plan was remarkable, not so much for the careful detail which any competent commanding general of the day might perhaps have conceived, but for the scale of planning and administration which set it in place and for the degree of coordination and communication which made it operational. The Greek artillery command structure would have had to be such that it was brought under one hand. Careful targeting and fire control would have been important but flexibility would have had to be retained, so as to be in a position to decentralize should the need arise for rapid, independent operations. Much would have rested upon a good system of communication. Training and rehearsal would have been an essential requirement and adequate quantities of weaponry would have had to be pre-sited, together with dumps of missiles of every variety, from ten talent stones, to stone-shot, to bolts and ammunition generally. The quantities in each case would have been huge and may perhaps be judged from the volume of war material seized by Scipio at New Carthage a few years later, where Livy[32] lists '120 catapults of the largest sort, 281 smaller ones; 23 large, 52 smaller *ballistae*; countless "scorpions" large and small, and a great quantity of equipment and missiles' as having fallen into his hands.

There has been much vigorous debate over the correct siting of such artillery weapons in a defensive siege role. It is linked with three essential requirements: namely, the need for an effective method of controlling range adjustments; the vital necessity of providing counter-battery fire, aimed at neutralizing enemy artillery directed at the destruction of the battlement defences or endeavouring to breach the curtain walls with their heavy stone-throwing engines; and the need to bring fire to bear upon the vulnerable blind spots, already mentioned above, at the base of curtain walls.

The range of non-torsion weaponry could only have been varied by increasing or lowering the angle of elevation. In later years, with the introduction of torsion weaponry, it was feasible to vary range by altering the tension of the spring but this also had the effect of lessening the force of the strike-impact and thus reducing the effectiveness of the weapon. It was not, therefore, a solution likely to render itself attractive to artillerymen. The height and presence of the wall was a key factor, adding to the effective range of artillery should this be based upon its summit; reducing the range, should the weapon be sited, albeit securely, behind the wall, and resolving many problems, including coverage of the blind spot, if positioned to its front. Philon of Byzantium, who lived towards the end of the third century BC and claimed he derived his knowledge of artillery matters from artificers in the arsenals of Rhodes and Alexandria, favoured the forward location.[33] He advised that platforms should be constructed in front of curtain walls and outworks for

> as many and as large engines as possible, some at surface level, others below ground level, so that there may be plenty of room for their operation, so that the detachments may not be hit, so that they may inflict casualties while themselves out of sight, and so that, when the enemy draws near, the aimers may not be handicapped by being unable to depress their engines.

The advantages of positioning artillery in this manner cannot be denied but it would necessarily have lain exposed and the possibility of losing vital weaponry to enemy attack would have been very real. Likewise, the ploy by Archimedes of positioning 'scorpion' catapults to fire through portholes in curtain walls, or even the suggestion that some machines might be sited within the walls themselves, could not under normal circumstances have been generally acceptable, for it surprised the Romans. The reason for this is probably that the wall would have been weakened by being adapted for this purpose and would, as a consequence, have been rendered vulnerable to battering-ram attack. Marsden, the author of authoritative works on Greek and Roman artillery,[34] is in no doubt that a fortification's defensive artillery was set up on the ordinary single rampart walk, as well as within the stalwart towers with which walls were provided from about the middle of the fourth century.

The use of siege artillery in an assault followed a broader pattern than defence, mainly because the besieger had greater room for manoeuvre. It was frequently found in a counter-battery as well as an anti-personnel role, where it repelled sorties and directed its fire at the defending garrison manning the battlements. On occasions heavy artillery, casting stone missiles of great weight, was used in an effort to breach the walls of a stronghold. At the siege of Jotapata, during the Galilean war, Josephus records that Vespasian fielded 160 artillery weapons, set them up in a ring and then ordered them to bombard the men on the wall in a 'synchronised barrage' of javelins, firebrands, showers of arrows and stones weighing nearly one hundredweight, 'whilst a host of Arab bowmen with all the javelin men and slingers let fly at the same time as the artillery'.[35] As at Syracuse, a barrage on this scale, synchronized with such a number and variety of weapons, must have called for a highly effective system of command and control: nor could it have been put into practical effect unless gun-crews had been well drilled in firing procedures.

Josephus, in his work *The Jewish War*, has provided us with many forceful descriptions of battle but probably none more lively than his account of the second phase of the assault by Titus upon the walls of Jerusalem. The Jews had captured a number of artillery weapons from the Romans when they had earlier chased Cestius out of Antonia. They had few men who understood how to operate them but were being instructed in their working by deserters and they were directing what fire they could generate at Roman engineers preparing a way for an assault with battering-rams. The legionary engines responding to this fire, wrote Josephus,

> were masterpieces of construction, but none were equal to those of the Tenth; their spear-throwers were more powerful and their stone-throwers bigger, so that they could repulse not only the sorties but also the fighters on the wall. The stone missiles weighed half a hundredweight and travelled four hundred yards or more; no one who got in their way, whether in the front line or far behind, remained standing. At first the Jews kept watch for the stone – it was white, so that not only was it heard whizzing through the air but its shining surface could easily be seen. Look-outs posted on the towers gave them warning every time a shot was fired from the engine and came hurtling towards them, by shouting 'Baby on the way!'. Those in its path at once scattered and fell prone, a precaution which resulted in the stone passing harmlessly through till it came to a stop. The Roman counter was to blacken the stone. As it could not then be seen so easily, they hit their target. . . .[36]

Under cover of this counter-battery fire the Romans moved forward their battering-rams, but were showered with missiles and firebrands, which set alight the lattice

Titus, son of Vespasian, emperor of Rome (AD 79–81) and victor of the siege of Jerusalem, AD 70

shields under which they were sheltering. Vespasian's son, Titus, the leader of this storming party, then placed 'the cavalry and bowmen either side of the engines, beat off the fire-throwers, repulsed those who were throwing missiles from the towers and got the battering-rams into action. Yet the wall did not give way. . . .'

Jerusalem, standing high upon a plateau and built upon a hard limestone formation (which, one must imagine, would have made the battlefield completion of defensive earthworks a heavy, arduous task, although Roman military discipline was to prove otherwise), provided a formidable obstacle to Roman ambitions. It was bounded on the west and south by the precipitous scarp of the Hinnom and on the east by the equally steep slopes of the valley of the Kidron, both of which rendered impossible the working of enemy battering-rams and assault towers. The tactical options for Titus, when making his plans, were limited. Moreover, within its walls, the city possessed an asset invaluable to its defenders and not available to its besiegers – a plentiful and secure water supply, said to have been completed by Hezekiah[37] when Jerusalem was under threat of siege by King Sennacherib of Assyria in 701 BC. This precious supply was conveyed by tunnel from the bountiful spring at Gihon, outside the south-eastern wall, to the Pool of Siloam, a reservoir securely located in the heart of the Upper and Lower Cities.

Apart from the natural strength of the terrain which surrounded it, Jerusalem, the 'stronghold of Zion', had powerful man-made defences. Initially built as a Jebusite town, it had fallen to the Israelites about 1000 BC and, according to 1 Samuel 5: 6–10, although relatively small compared with the scale of its later development, it had subsequently been powerfully fortified by David. A part of the original Jebusite city wall, repaired and strengthened in those years and recently revealed by archaeological excavation, measured a doughty 27 feet wide.

In the intervening years between its capture by the Israelites and the arrival of the Romans, Jerusalem grew in size and expanded in the only direction open to it, across the plateau, northwards and to the west. At various stages of its development, further substantial additions to its defences were made. Firstly, a northern wall, which reached eastwards from Herod's Palace, by the Jaffa Gate, to the Temple, was built to embrace the Upper City. Later, a further, second wall, was added by Antipater, the father of Herod Agrippa I. This pursued a zig-zag course from the palace to the Antonia: it was given added strength by the provision of a large fortification, the Middle Tower, which guarded its gate and acted as the hinge of its defence. The Antonia had been Herod's earlier home before he provided himself with a grander residence. Finally, a third wall was constructed by Agrippa I, king of Judaea (AD 41–44) but he failed to finish it. He feared that the added strength he was giving to the city might be misconstrued by his Roman masters as a gesture of defiance. Thus, at the moment of Titus's arrival, a portion of it, the return wall lying between the octagonal tower of Psephinus and the western, Jaffa Gate, still stood incomplete. Apart from this, it was a magnificent structure, 15 feet wide and 40 feet tall.

The defenders of the city, if one accepts Josephus' assessment of their quality, were a strangely mixed bag of some 23,400 men who had imposed themselves upon a reluctant population. In the main they were private armies, thrown up during Vespasian's earlier operations in Galilee. They arrived piecemeal and quickly

PHASE I

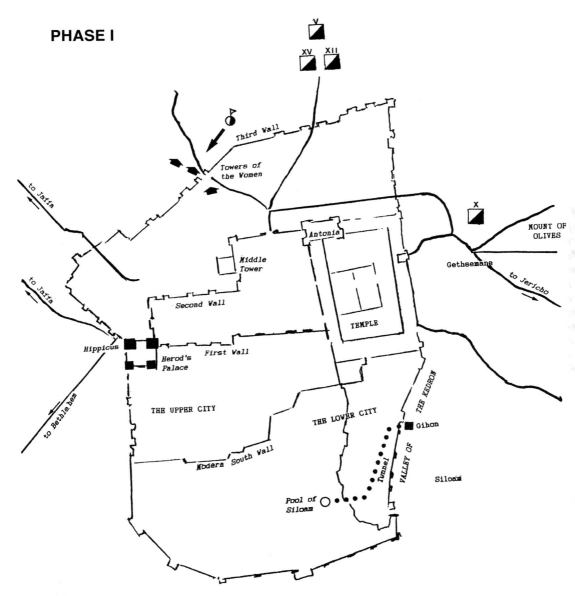

Titus, upon arrival, encamped his legions to the north of the city and deployed X Legion on the Mount of Olives, confronting its eastern walls. Meantime, he set forth to conduct a reconnaissance of the partially unfinished wall adjoining the Jaffa Gate. Here he was ambushed. Note the city's secure water supply flowing underground from Gihon to the Pool of Siloam

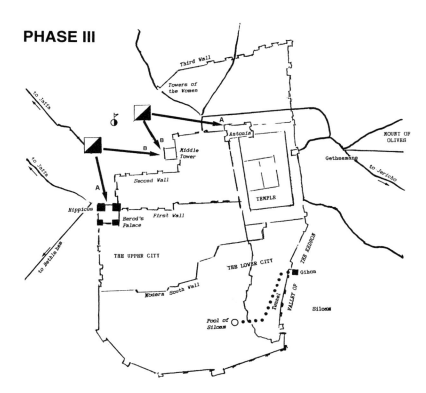

PHASE III

Titus opened the second phase of his siege of the city by grouping his army in two divisions and assaulting the wall running north-west from Herod's Palace. When this had been overcome, he directed his assault (Phase III, illustrated here), first, against the extremities of the second wall (arrows A) and then against the Middle Tower

divided themselves into two factions, treating each other with an uninhibited distrust which frequently erupted into open hostility. John of Gischala, leader of 6,000 Galileans and described by Josephus as a 'gangster and racketeer', occupied the Antonia and the Temple: here, he was joined by 2,400 Zealots, extreme patriots with a reputation for brutality, guided primarily by religious motives. The second faction was led by Simon bar Gioras, who occupied the City with its shops, magazines and foodstores. It comprised 10,000 Sicarii, fervent nationalists who lived by the dagger from which they derived their name, and 5,000 Idumaeans who had defected to his side during his recent invasion of their territory. During the siege warfare which was about to open, the two parties reconciled their differences and joined forces to great effect. During a lull in the fighting, however, they turned upon each other once again and destroyed a large part of their food resources in a senseless, fanatical encounter. According to Josephus, 'all the environs of the Temple were reduced to ashes, the city was converted into a desolate no-man's-land for their domestic warfare, and almost all the corn, which might have sufficed them for many years of siege, was burnt up'.[38]

These were the defences and the garrison which opposed Titus upon his arrival.

The Roman general had assembled an army of some 65,000 men. This was made up of four legions: the Vth from Emmaus; the Xth, of high reputation, had marched from Jericho; the XIIth, which was seeking to redeem its name after its recent ignominious defeat under Cestius Gallus; and the XVth, Titus's own

PHASE IV

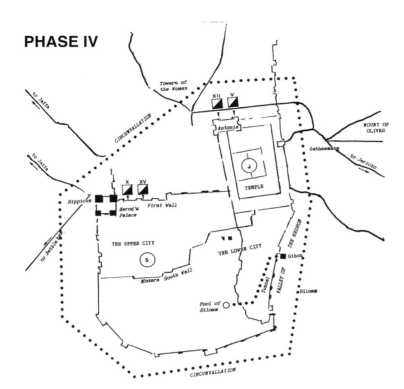

In this final phase Titus completed the destruction of the Second and Third walls and simultaneously assaulted the defences at the extremities of the First wall. Note the circumvallation with which he surrounded the city at this stage in order to contain the enemy within its walls and cut their supply lines

regiment which had come up from Alexandria, together with 1,000 men each from the IIIrd and XXIInd Legions, and a further 3,000 from the Euphrates garrisons. The balance of his force, twenty allied cohorts, eight regiments of cavalry, with a variety of auxiliaries armed with supporting weaponry, was furnished by Agrippa II and the rulers of neighbouring client kingdoms.

Titus was careful not to disclose his hand too soon. He approached Jerusalem from the north and camped in a depression 4 miles distant, north of Beit Hanina, invisible to look-outs posted high on the city's fortifications. From there, he set out on a reconnaissance with a bodyguard of some 600 men and, riding through the maze of vegetable gardens and partitioning walls which covered the ground on that side of the city and provided a source of food for it, he made his way directly towards the Damascus Gate. Here he turned right and headed in the direction of the octagonal tower of Psephinus but, as he passed the Towers of the Women, the Jews streamed out in considerable force and ambushed him. He managed to fight his way free. He now ordered his legions forward to new positions, the Xth to occupy the Mount of Olives, opposite the Golden Gate and the remainder to Mount Scopus; and, when they had completed the fortification of their camps, they were instructed to begin the task of levelling the ground between their position and the Jaffa Gate. He wanted no further surprises.

Titus would almost certainly have been aware of the unfinished state of the wall beside the Jaffa Gate when he rode forward to reconnoitre it. He quickly recognized its weakness and at once re-assembled his entire force, in two army groups, with battering-rams supported by spearmen, bowmen and slingers, to attack the reach between Herod's Palace and the Psephinus Tower. The Jewish defenders resisted stubbornly but found themselves unable to counter the fire which the Romans, in support of their battering-rams, brought to bear upon them from their 75 feet high assault towers. On their lofty perch, explained Josephus,

these men were beyond the reach of Jewish weapons and there was no way to capture the towers . . . so they withdrew out of range, abandoning the attempts to hold off the assaults of the Romans, which by their incessant blows were little by little effecting their purpose. . . . Besides, through laziness and their habit of deciding wrongly, they thought it a waste of effort to defend this wall as there remained two more behind it. Most of them slacked off and retired: and when the Romans climbed through the breach made by Victor (the battering-ram) they all left their posts and ran helter-skelter to the second wall.[39]

This phase of the siege had taken Titus five days to complete. He now issued orders for the destruction of the outer wall (or Third Wall, as it is known), in its entirety from the Jaffa Gate to the north-eastern corner of the city. When this had been done, he commenced his attack upon the next line of defence, which now confronted him, the Second Wall.

Here, he determined to stretch the Jewish defences by launching his two army groups against it simultaneously, one to assault the western end of the wall, defended by Simon, adjoining Herod's Palace, and the other against its eastern extremity, the Antonia and the colonnade of the Temple occupied by John. He soon appreciated, however, that this tactic played to his enemy's advantage for, by dividing his forces in this manner he ensured that his troops got little rest, whereas the restricted width of the wall limited the numbers which the Jews needed to field at any one time. As a consequence, he now determined to concentrate his assault upon the Middle Tower, attacking in waves so that, while some of his legions were engaged in the arduous, dangerous work of fighting to penetrate the walls, others could be resting. He thus always had fresh formations to bring to the task. In this manner, after a sustained and bitter struggle, he broke through on 25 May,[40] fifteen days after his success in breaching the outer wall, but it was not until 4 June that he consolidated his hold upon it. He occupied the intervening days by holding a ceremonial pay parade, in the hope that this glimpse of Roman wealth and resources might persuade his enemy to yield. It did not.

Titus now embarked upon the final and, as it was to prove, the most lengthy and bloody stage of the siege, which was to endure until 26 September, when the last resistance within the city was overcome. First, he set about the systematic destruction of the newly breached Second Wall, with the exception of the reach on the extreme west, where it ran north and south. He retained this as a protective screen for his new camp and then set the Vth and XIIth Legions to work, facing the Antonia, building parallel ramps, 30 feet apart, upon which to site his assault towers. He gave a similar task to the Xth and XVth Legions, at the western extremity of the wall, north of the Upper City. These platforms were of vast size

Siege towers at Jerusalem, AD 70 (Polybius, ed. 1727)

and took seventeen days of intense labour to complete. As soon as they were finished, the great assault towers were moved forward to them in preparation for the next stage. The Romans, however, were unaware that John of Gischala had not been idle. As they laboured opposite the Antonia, so also had he been working hard, undermining their work with tunnels and caves, stuffing these with faggots daubed with pitch and bitumen, primed to be set alight. He chose his moment carefully before igniting them and then, as they turned to ashes in the heat of the flames they generated, the platforms fell with a thunderous crash into the cavity thus created:

> At once there was a dense cloud of smoke and dust as the flames were choked with debris; then when the mass of timber was burnt away a brilliant flame burnt through. This sudden blow filled the Romans with consternation, and the ingenuity of the Jews filled them with despondency; as they had felt sure that victory was imminent, the shock froze their hope of success even in the future. To fight the flames was useless, for even if they did put them out their platforms were already swallowed up.[41]

But worse was to come: two days later, Simon's forces emerged from behind their defences in a surprise raid upon the two platforms confronting his section of the wall. These they also set alight and doggedly refused to be driven away until the structures were well ablaze.

As a result of these counter-strokes, Titus was confronted by a crisis of resources. He had used his last remaining stocks of timber in the construction of the siege

Captured spoils from Jerusalem carried in Titus's triumphal procession, c. AD 81

platforms, and replacements were unavailable. The time, moreover, was approaching mid-summer and, in the desert sun, he was running short of water. Josephus makes no mention of Titus's administrative supply lines but the Roman had, by this time, already been positioned in front of Jerusalem for some five to six weeks. In this circumstance, he must already have been well into a programme of re-supply, hauling grain and water for men and animals from some distance away. The Jews, we may rest assured, would not have allowed these vital columns to move unscathed, for they appear to have had many means of penetrating under and over their city walls. Titus, himself aware of his enemy's self-created shortage of rations, now determined to seal off this traffic by constructing a circumvallation around Jerusalem, in a circular, 5 mile long fortification, strengthened at intervals by thirteen towers, each some 200 feet in circumference, between which his army constantly patrolled, night and day. In the end, before their surrender, the besieged were reduced to eating grass and the leather of their belts and sandals.

Titus's siege of Jerusalem provides a classic example of the art of siege warfares: the effectiveness of the countermining operations by the besieged, the importance to both sides of carefully planned administration and the need for plentiful resources. Particularly noteworthy, on the Roman side, is the manner in which the Roman general tightened his grip on his newly acquired territory at the successful end of each of his assault phases, before moving to the next, thus denying his enemy any opportunity of re-occupying their old defences by counter-attack. The strength of the great Roman war machine might have been unavailing, however, had it not been for the internecine fighting among their enemy that destroyed their stocks of food. Had this not happened, by their ingenious and heroic destruction of Titus's siege platforms, the Jews might have achieved a stunning victory.

Appendix 1

THE PEUTINGER MAP

This map – a road map of the Roman Empire – takes its name from the German map collector Conrad Peutinger (1465–1547). Peutinger received this map, a manuscript of the twelfth or thirteenth century, from Conrad Celtes, and copied only two sections of it. The section illustrated here (p. x) is based on the 1653 edition by Jan Janson.

It is conjectured that the Celtes manuscript was copied from an ancient map drawn around 365 CE. It describes the roads of the Roman Empire from Gaul in the west to India in the east. The map takes the form of a long, narrow parchment scroll (approximately 33 × 682 cm), divided into twelve sections. Section one, the extreme west, was lost. The Land of Israel appears in section eleven. There, as in the rest of the map, important settlements are noted, with distances between them marked in Roman miles.

Jerusalem is depicted by two buildings and the Mount of Olives. Above it appears the inscription: 'Formerly called Jerusalem (and today) Aelia Capitolina'. One of the roads leaving Jerusalem crosses the Negev Desert and ends at Eilat on the shores of the Red Sea. Among other sites mentioned on the map are the Jordan River, Tiberias, the Dead Sea, Jaffa, Ptolemaida ('Akko), Azoton (Ashdod) and Ascalon (Ashqelon).

In an attempt to accommodate the entire empire in a narrow scroll, the cartographer had to make the map narrow. To do so he reduced the areas of the sea to narrow strips. As a result, many lines running north to south are sketched in the map from right to left. The proper dimensions for east to west are largely maintained. The Peutinger Map probably served travellers to the Land of Israel, including pilgrims.

A story is told about the Roman road from Jerusalem to Eilat to the effect that, during the War of Independence in 1948, the Israeli High Command had a secret plan to conquer the Negev and Eilat. This plan was called 'Operation Fact'. Yigael Yadin, the chief of operations and an archaeologist by profession, knew of the Roman road leading from Jerusalem to Eilat – shown in the Peutinger Map – parts of which had been preserved under the desert sand.

The armoured column sent to conquer Eilat followed this route and reached the shores of the Red Sea. In doing so, they caught the enemy by surprise, for the Jordanians did not believe that the Israeli army would succeed in crossing the desert in the absence of a partly paved road.

Appendix 2

Signal Methods

I Commands by Voice, Bugle and Signal

The commands must be short and unambiguous. This would be attained if the particular command should precede the general, since the general are ambiguous. For example, we would not say, 'Face right!' but 'Right Face!', so that in their eagerness some may not make the turn to the right and others to the left when the order to turn has been given first, but that all may do the same thing together; nor do we say 'Face about right!' but 'Right about face!', nor 'Counter march, Laconian!' but 'Laconian, counter march!' and [passage missing]. . . .

Stand by to take up arms! Baggage men fall out! Silence in the ranks! and Attention! Take up arms! Shoulder arms! Take distance! Shoulder spear! Dress files! Dress ranks! Dress files by the file-leader! File-closer, dress file! Keep your original distance! Right Face! Forward march! Halt! Depth Double! As you were! Depth half! As you were! Length double! As you were! Laconian, counter march! As you were! Quarter turn! As you were! Right half turn! As you were!

These are in brief the principles of the tactician; they mean safety to those who follow them and danger to those who disobey.

Ascelepiodotus, *Tactics*, 11

II Advanced Techniques

There is no doubt that the Romans practised several advanced signalling methods, many of the more ingenious being inherited from the Greeks. Vegetius[1] describes a sophisticated semaphore system (another type of semaphore was employed by the Roman navy – see Chapter 6, pp. 115–17), operated from signal towers and equipped with wooden arms. Its range would necessarily have been minimal without the benefit of the telescope provided to the signal stations which, in the nineteenth century, linked naval dockyards on the south coast of England with the Admiralty building in Whitehall. These were sited, spaced every 7 or 8 miles, on suitable vantage points along the way, sometimes at the top of great buildings such as the Duke of York's headquarters in Chelsea: but if a message were to be recorded by the naked eye, as would have been the case in Roman times, the intervening distance would have needed to be much less.

Polybius has recorded two techniques, both of them highly ingenious.[2] The first he attributed to Aeneas, 'the author of the work on strategy'. It required the possession of two completely identical earthenware jars, each 3 cubits (or roughly 5 feet) in depth and 1 cubit (or 20 inches) wide, each drilled, low down, with apertures of similar size and then plugged; and each provided with a cork a little less in diameter than the mouth of the jar. A light rod, calibrated in equal sections of three finger-breadths, was driven through the exact centre of each cork so that, when the jars were ultimately filled with equal levels of water, they floated at the same height. The calibrations on the rods were marked, the one against the other, with carefully chosen and identical phrases, listed in an identical sequence. Thus, when both plugs were removed simultaneously and the water flowed out, the rods sank at an equal speed to an equal level.

The transmission of the message was simply achieved. The despatching signaller had first to display a lighted torch from his signal station and then wait until the distant recipient had raised another in response. When this had been done, and both were clearly visible, the despatcher lowered his torch, at which moment other signallers, at both the despatching and receiving ends, extracted the corks from their respective jars and allowed the water to escape until the message inscribed on the rod was level with the lip of the jar. At this moment, the despatcher's torch was again displayed and both corks were promptly replaced. If the signal drill had been carried out with dexterity and speed, then the rods should have fallen identical distances, enabling the message to be passed.

Polybius opined that this method was an improvement on the hilltop beacon, lit in accordance with a 'preconcerted code', but he argued that both methods had their obvious limitations, for

> . . . it is impossible to agree beforehand about things of which one cannot be aware before they happen. And this is the vital matter; for how can anyone consider how to render assistance if he does not know how many of the enemy have arrived, or where? And how can anyone be of good cheer or the reverse, or in fact think of it at all, if he does not understand how many ships or how much corn has arrived from the allies?

For these reasons, Polybius expressed his preference for a later method, 'devised by Cleoxenus and Democleitus and perfected by myself'. In this case, the alphabet was divided into five roughly equal parts and listed upon five 'tablets', numbered one to five. Each letter on each tablet was also allocated a number and these were arrayed in front of the despatching signaller, so that he might read them with ease. Under his command were two groups of transmitting signallers, each group stationed behind a screen, one on his left, the other on his right. Each team held at least six fire torches ready to hand. The purpose of the screens was to conceal their light when not in use.

When the despatcher wished to send a message, he raised two torches to attract the attention of the intended recipient who was required to respond by raising two torches in reply. Both sets of torches were then withdrawn. The message was now passed by spelling out the individual letters of each word, the number of torches exposed over the left-hand screen indicating the number of the tablet to be used, the number of torches over the right-hand screen indicating the number of the letter on the specified tablet. Those engaged in the work, stressed Polybius, 'must have proper practice so that, when it comes to putting it in action, they may communicate with each other without the possibility of a mistake'. In order to be certain the sentries did not waste time, or miss an incoming message, each signal station was required to plant a hollow tube, sighted on a fixed line and directed towards its neighbouring post.

What degree of efficiency could this system have achieved? Polybius, in an ambiguous phrase, states that the methods described by him were capable of despatching every kind of urgent message over 'distances of three, four or even more days' journey'. If one accepts a day's journey to be 20 miles then, at first sight, it appears he is mooting a single span of 100 miles or more, a clearly inconceivable range: more probably, he is suggesting the total distance across which a message could have been efficiently and economically relayed. On the Dacian frontier, for example, a system of Roman signal stations, watch towers and forts, all located about 6 miles apart, has been found between Bologa and Tehau.[3] These could all have played some part in a local signal network of this nature but, broadly, this still requires confirmation. Where, when and for what purpose such systems would have been employed are questions which still await answers.

1.   Vegetius, *Epitoma rei militaris*, pp. iii, 5.
2.   Polybius, *The Rise of the Roman Empire* (Penguin Classics, London, 1979), pp. x, 43–7.
3.   See Nicolae Gudea, 'The Defensive System of Roman Dacia', *Britannia* (1979), Vol 10.

Appendix 3

MARCHING-CAMP TECHNIQUES

Labour Figures for Common Engineering Tasks

Serial No.	Description	Unit	Output per man per hour	Notes
(a)	(b)	(c)	(d)	(e)
	Site clearance and demolition work			
1	Clearing bushes and scrub	m²	33	
2	Clearing dense undergrowth with saplings up to 100 mm diam.	m²	11.7	
	Earthwork			
1	Digging to depth not exceeding 1 m including loading into barrows	m³	0.4–0.7	(i) (ii)
2	Digging only, to depth not exceeding 1.5 m and throw not exceeding 2 m	m³	0.4–0.7	(i) (ii)
3	Shovelling loose soil, throw not exceeding 3 m	m³	1.1–1.5	(i) (ii)
4	Filling loose soil into sandbags	bag	20	(i) (iii)
5	Filling loose soil into barrows	m³	1.1–1.5	(i)
6	Filling wet mud into barrows	m³	1.3	(i)
7	Wheeling 25 m, depositing and returning empty	m³	2.5–3.0	(i) (iii)
8	Spreading soil in 150 mm layers	m³	1.5–2.0	(i) (ii)
9	Ramming soil in 150 mm layers	m²	1.5–2.0	(i) (ii)
10	Levelling and trimming slopes to profile	m²	8	

(i) Daylight rate. For night with no moon: x½. For night with full moon or artificial moonlight: × 2/3

(ii) Depends on nature of soil (chalk to sandy loam)

(iii) Each metre rise is equivalent to 6 m on level; maximum slope 1 in 8

Note: statistics provided by The Royal School of Military Engineering and are MOD copyright.

Appendix 4

SIEGE OF JERUSALEM, AD 70

Sequence of Main Events

 4 May   Titus arrives with his army, encamps and then carries out reconnaissance of city walls
10 May   Siege opens
25 May   Romans overwhelm the Third (outer) Wall, raze the NW reach to the ground and occupy the New City.
16 June   Roman ramps (platforms) at Antonia undermined by John of Gischala
18 June   Simon destroys the platforms confronting his stretch of the wall
14 July   Titus renews assault on Antonia
24 July   Standard-bearer of the Vth Legion, with trumpeter, gains a foothold on the wall. Jewish defenders seek refuge in the Temple
29 Aug.   Temple falls and contents looted; defenders now pull back to Upper City
 8 Sep.   Titus recoups timber stores from source 10 miles away and constructs two new platforms over next 18 days
18 Sep.   Jerusalem capitulates after 139 days of siege

Notes

Introduction

1. Major General J.F.C Fuller, *Julius Caesar, Man, Soldier and Tyrant* (London, 1965), pp. 315–6.
2. Chester Wilmot, *The Struggle for Europe* (London, 1952), p. 75, fn 2.
3. Field Marshal Sir William Slim, *Defeat into Victory* (London, 1956), pp. xxi, 485.
4. Julius Caesar, *de Bello Africo*, 8.
5. Josephus, *The Jewish War*, Ch. III, 109.
6. Vegetius, *Epitoma rei militaris*, I, p. 1.
7. John Peddie, *Invasion, The Roman Conquest of Britain* (Gloucester, 1987), Ch. III.
8. Field Marshal Viscount Montgomery, *Concise History of Warfare*, I, p. 19.
9. Onasander, *The General*, Ch. III.
10. Livy, xxxi, 42.
11. *ibid*; xliv, 37.
12. Vegetius, *op. cit.*, II, 31.

Chapter One

1. Field Marshal Sir William Slim, *Defeat into Victory* (London, 1956), IX, p. 186.
2. Vegetius, *Epitoma rei militaris*, I, p. 12.
3. J.A. Froude, *Caesar, A Sketch* (London, 1886), p. 550.
4. Theodor Mommsen, *The History of Rome* (London, 1911), IV, p. 430.
5. Dr T. Rice Holmes, *Caesar's Conquest of Gaul* (Oxford, 1911), pp. xii, 42.
6. Professor H. Last, *Cambridge Ancient History* (Cambridge, 1932), IX, p. 705.
7. T. Dodge, *Caesar, Great Captains Series* (Boston, Mass. 1892), Vol. 11, p. 767.
8. *ibid.*, pp. 692–3.
9. J.F.C. Fuller, *Julius Caesar, Man, Soldier and Tyrant*, (London, 1965), p. 318.
10. In AD 44, Caesar's dictatorship, previously bestowed for only ten years, was declared perpetual, an act which almost certainly contributed to his assassination.
11. W.V. Harris, *War and Imperialism in Republican Rome*, 327–70 BC (Oxford, 1979), pp. 13–17.
12. Pliny, *Epistulae*, VIII, 14, 4–5.
13. Cassius Dio, *The Roman History*, lx, 19; the official line from Rome was that Plautius had been ordered to notify Claudius if and when his invasion force got into difficulties; that Plautius was stalled in his efforts to cross the Thames and sent for the emperor, who duly arrived and resolved his general's problems. However, Claudius could not have afforded to be associated with failure, and we may therefore judge that his victory was already assured when he sailed upriver – indeed, when he left Rome.
14. Cicero, *Ad Familiares*, XV.
15. Frontinus, *Stratagems*, IV, vii, 5.
16. Tacitus, *The Agricola*, 20.
17. Velleius Paterculus, II, lxvii, 6, 1.
18. Suetonius, *The Twelve Caesars*, 60.
19. Caesar, *de Bello Gallico*, IV, 14; VII, 56; VIII, 3.
20. Caesar, *de Bello Africo*, 26.
21. Fuller, *op. cit.*, p. 322.
22. Frontinus, *op. cit.*, IV, 1, 7.
23. Velleius Paterculus II, 114, 1–2, as in R.W. Davies, 'The Roman Military Medical Service', *Saalburg Jahrbuch* (1970), p. 98, fn 110.
24. Tacitus, *Annals*, 1, 69.
25. For a full account see Davies, *op. cit.*
26. *de Bello Gallico*, VI, 36.
27. Montgomery, *Concise History of Warfare* (London, 1972), pp.1, 17.
28. *de Bello Gallico*, VI, 7.
29. Frontinus, *op. cit.*, II, p. viii.
30. Vegetius, *Epitoma rei militaris*, I, pp. 1, 9.
31. Tacitus, *op. cit.*, xiii, 34.

32. Frontinus, *op. cit.*, IV, 1, 1–46.
33. *ibid.*, IV, i, 38.
34. Suetonius, *op. cit.*, 1, 62.
35. *ibid.*, 1, 68.
36. Frontinus, *op. cit.*, III, xiv, 2.
37. Vegetius, *op. cit.*, III, p. 3.
38. Frontinus, *op. cit.*, IV, pp. iii, 14.
39. Tacitus, *op. cit.*, xii, 29.
40. Tacitus, *The Agricola*, 21.
41. Frontinus, *op. cit.*, III, xvi; I, ix.
42. Suetonius, *op. cit.*, I, 65.
43. Onasander, *The General*, III.

Chapter Two

1. Polybius, *The Rise of the Roman Empire* (Penguin Classics, London, 1979), p. ix.
2. Asclepiodotus, *Tactics*, p. 10.
3. Vegetius, *Epitoma rei militaris*, p. ii.
4. Homer, *Iliad*, 18.
5. Curt Sachs, *The History of Musical Instruments* (London, 1940), p. 145.
6. Michael P. Speidel, *Eagle Bearer and Trumpeter* (Rheinisches Landesmuseum, Bonn, 1967), p. 154.
7. *ibid.*
8. Philip Bate, *The Trumpet and the Trombone* (London, 1978), pp. 5, 101–5.
9. Speidel, *op. cit.*, p. 161.
10. Sir Ian Richmond, *Trajan's Army on Trajan's Column* (British School at Rome London, 1982), p. 49 and pl. 20a.
11. Polybius, *op. cit.*, vi, 40; and Josephus, *The Jewish War* (Penguin Classics, London, repr. 1985), iii, 87.
12. *Regulations for the Exercise of Riflemen* (The War Office, London, 1801).
13. Caesar, *War Commentaries*, ed. John Warrington (London, repr. 1965), III, pp. 292–4.
14. Frontinus, *Strategems* (Loeb Classical Library, London, repr. 1980), pp. 121.
15. Vegetius, *op. cit.*, III, 'Proper Distances and Intervals'.
16. Dr H.G. Farmer, *The Rise and Development of Military Music* (London, 1912), pp. i, 10.
17. See, among others, Lawrence Keppie, *The Making of the Roman Army* (London, 1984) and Graham Webster, *The Roman Imperial Army* (London, 2nd edn. 1979).
18. Field Marshal Viscount Montgomery, *A Concise History of Warfare* (London, 1972), pp. iv, 53.
19. Caesar, *de Bello Gallico*, I, 39, 2.
20. Polybius, *op. cit.*, vi, 24.
21. Vegetius, *op. cit.*, I, Prefect of camp, p.16.
22. For example, G. Webster, *The Roman Army* (Chester, 1956), II; George C. Boon and Colin Williams, *Plan of Caerleon* (Cardiff, 1967), p. 6.
23. Speidel, *op. cit.*, 138.
24. Caesar, *op. cit.*, V, 37.
25. Tacitus, *The Annals of Imperial Rome* (Penguin Classics, Harmondsworth, 1981), II, 14.
26. As in Graham Webster, *The Roman Imperial Army* (London, 1979), pp. 3, 137 fn 2 (Florus, pp. iv, 12).
27. Vegetius, *op. cit.*, Centuries and Ensigns of the Foot, II, 50.
28. Polybius, *op. cit.*, vi, 24.
29. Dio, *Roman History*, III, xl, 18.
30. Caesar, *de Bello Africo*, 15.
31. Josephus, *The Jewish War* (Penguin Classics, Harmondsworth, 1987), II, 587.
32. *ibid.*, III, 111.
33. Caesar, *de Bello Africo*, 17.
34. A.M. Ramsay, The Speed of the Roman Imperial Post, *Journal of Roman Studies*, pp. 60–1.
35. Suetonius, *Augustus*, 49–50.
36. As in Ramsay, *op. cit.*, p. 67.
37. Plutarch, *Julius*, 17.
38. *ibid.*, 68.
39. Procopius, *Anecdota*, 30.
40. Suetonius, *Claudius*, 17.
41. Procopius, *op. cit.*

Chapter Three

1. Vegetius, *Epitoma rei militaris*, pp. iii, 71.
2. Chester Wilmot, *The Struggle for Europe* (London, 1952), pp. xxiv, 470.
3. J.F.C. Fuller, *Julius Caesar, Man, Soldier and Tyrant* (London, 1965), pp. xiv, 316.
4. Caesar, *The Conquest of Gaul* (Penguin Classics, Harmondsworth, 1984), vii, 14.
5. *ibid.*, v, 45.
6. I.A. Richmond and J. McIntyre, 'Tents of the Roman army and leather from Birdoswald', *Transactions of the Cumberland and Westmorland Archaeological Society* (1934), p. 34.
7. *op. cit.*, vi, 36–7.
8. Donald W. Engels, *Alexander the Great and the Logistics of the Macedonian Army* (Berkeley, 1978), pp. i, 15.
9. *ibid.*, pp. i, 15, fn 15.
10. Generally calculated at 3 to 3½ miles in the hour, including a ten minute rest period.

11. They were, of course, on call in case of emergency and this, as will be seen, carried its risks.
12. Fuller, *op. cit.*, pp. xiv, 317.
13. Sir Ian Richmond, *Trajan's Army on Trajan's Column* (London, 1982), pp. i, 12–13.
14. Livy, *The Early History of Rome*, iii, 27.
15. On the basis that each wagon carries 2 tons.
16. Patrick Boyle and James Musgrave-Wood, *Jungle, Jungle, Little Chindit* (London, 1944) pp. 4, 22.
17. Vegetius, *op. cit.*, pp. iii, 79.
18. Caesar, *op. cit.*, ii, 24.
19. That is, one servant for each of the 'Roman' cavalry and none for the native cavalry; but one servant/animal handler per infantry section. See also Engels, *op. cit.*
20. The GD drivers would have been employed transporting 'bulk' supplies, e.g. grain reserves, under the direct control of the baggage master and with no loyalties in any other direction.
21. Ann Hyland, *Equus, The Horse in the Roman World* (London, 1990), pp. 6, 87–100.
22. Caesar, *op. cit.*, i, 16.
23. *ibid.*, i, 11.
24. *ibid.*, i, 16.
25. *ibid.*, vii, 34.
26. The transport opportunities offered by these and other French riverways are examined in some detail in Chapter 6, pp. 106–9.
27. T. Pakenham, *The Boer War* (London, 1979), pp. 379–80.
28. Caesar, *op. cit.*, vii, 67.

Chapter Four

1. *Design for Military Operations – The British Military Doctrine* (prepared under the direction of the Chief of General Staff, 1989).
2. Frontinus, *Stratagems and Aqueducts*, ed. G.P. Gould (Loeb Classical Library, London, 1980), IV, i, 14.
3. General Carl Von Clausewitz, *Principles of War* (London, 1943) 3, pp. iii, 47.
4. See, e.g., Caesar, *The Conquest of Gaul* (Penguin Classics, Harmondsworth, 1984) vi, 32, where Caesar recounts the re-occupation of an old fortification.
5. *ibid.*, II, 20.
6. Major General J.F.C. Fuller, *Julius Caesar,*

Man, Soldier and Tyrant (London, 1965), pp. iv, 87.
7. Caesar, *op. cit.*, IV, 86–7.
8. Fuller, *op. cit.*, IV, 76.
9. Polybius, *The Rise of the Roman Empire* (Penguin Classics, London, 1979), VI, 27–34.
10. Vegetius, pp. iii, 82–3.
11. *ibid.*
12. Josephus, *The Jewish War* (Penguin Classics, London, repr. 1985), III, 87.
13. Polybius, *op. cit.*, VI, 41.
14. Josephus, *op. cit.*, III, 110.
15. Tacitus, *The Annals of Imperial Rome* (Penguin Classics, London, repr. 1981), I, 51.
16. Caesar, *op. cit.*, V, pp. 49, 5.
17. *ibid.*, II, 17–27.
18. Sir Ian Richmond, *Trajan's Army on Trajan's Column* (London, 1982), pp. i, 11–13.
19. Caesar, *op. cit.*, V, 33.
20. *ibid.*, VI, 5.
21. *ibid.*, VII, 18.
22. *ibid.*, III, 24.
23. *ibid.*, II, 4.
24. *ibid.*, 19.
25. Richmond, *op. cit.*
26. Caesar, *op. cit.*, II, 19.
27. *ibid.*, 26.
28. *ibid.*, 16.
29. Josephus, *op. cit.*, III, 134.
30. I have confirmed this spacing with the regimental sergeant-major at the School of Infantry. The fact that the men were carrying shields and javelins should have made little difference.
31. Vegetius, *op. cit.*, I, p. 30.
32. Caesar, *op. cit.*, VII, 40–1.
33. Ann Hyland, *Equus, The Horse in the Roman World* (London, 1990), pp. vi, 90.
34. David J. Breeze, 'The Logistics of Agricola's Final Campaign', *Talanta*, 16–19 (1987/8), pp. 7–22.
35. Vegetius, *op. cit.*, III, p. 82.
36. *ibid.*, p. 84.
37. Polybius, *op. cit.*, VI, 28–35.
38. S.S. Frere and J.K.S. St Joseph, *Roman Britain from the Air* (Cambridge, 1983) pp. ii, 23–4.
39. Polybius, *op. cit.*, VI, 31; on the other hand, Hyginus, writing late in the second century, records the space as 60 feet.
40. Caesar, *op. cit.*, VI, 37.
41. Arthur Feller, *The Fall of the Roman Empire:*

The Military Explanation (London, 1986), pp. ii, 28.
42. Vegetius, *op. cit.*, I, p. 13.

Chapter Five

1. Vegetius, *Epitoma rei militaris*, III, p. 93.
2. We may judge that this refers to both light (*ballista*) and heavy (*onager*) artillery, but particularly the latter.
3. Sir Ian Richmond, *Trajan's Army on Trajan's Column* (London, 1982), pp. 2, 19, fn 22.
4. Vegetius, *op. cit.*, I, p. 23.
5. W.W. Tarn, *Hellenistic Military and Naval Developments* (Cambridge, 1930), I, p. 20.
6. Livy, *Rome and the Mediterranean*, xxxviii, 21.
7. Vegetius, *op. cit.*, II, p. 58.
8. *ibid.*, pp. 52, 58.
9. C.T. Lewis and C. Short, *Freund's Latin Dictionary* (Oxford, 1879).
10. Caesar, *de Bello Gallico*, II, 7: I am indebted to my friend Colonel Charles Lane for the information that, when serving in the Oman, he witnessed Arabs knocking over small game with slingshot at ranges between 30 and 50 yards.
11. The authorship of the works which deal with Julius Caesar's Alexandrian, African and Spanish Wars is a matter of uncertainty. It is sometimes attributed to one Hirtius, a comparatively junior officer with limited access to the inner counsels of his commander-in-chief. Most scholars are inclined to accept that the true identity of the author remains obscure.
12. Caesar, *de Bello Africo*, 27.
13. Polybius, *The Rise of the Roman Empire* (Penguin Classics, London, 1979), III, 42–7.
14. Major H.G. Eady, R.E., The Tank, *United Services Journal* (1926), p. 81.
15. Montgomery of Alamein, *A Concise History of Warfare* (1972), pp. 18, 291–2.
16. Caesar, *de Bello Africo*, 83.
17. *ibid.*, 83.
18. Richmond, *op. cit.*, pp. 2, 19, fn 22.
19. Caesar, *de Bello Africo*, p. 19.
20. Richmond, *op. cit.*, pp. 17–20, Plate 4.
21. Tacitus, *The Germania*, 45.
22. Sir Ralph Payne-Gallwey, *The Projectile-Throwing Engines of the Ancients* (London, 1907; repr. 1973); 'Treatise on the Turkish and other Oriental Bows'.

23. Robert Hardy, *Longbow* (Cambridge, 1976), pp. 14–21.
24. Payne-Gallwey, *op. cit.*, Introduction, p. vii.
25. *ibid.*
26. Caesar, *de Bello Africo*, 12–14.
27. *ibid.*, 34.
28. Frontinus, *Stratagems*, II, ii, 5; this tactic by Ventidius would appear to confirm the range of a war arrow estimated by Payne-Gallwey (fn 22 above) as 360–400 yds.
29. Josephus, *The Jewish War*, III, 60.
30. Caesar, *de Bello Africo*, 60.
31. *ibid.*, 78.
32. Caesar, *de Bello Gallico*, II, 19.
33. Josephus, *op. cit.*, III, 220.
34. *ibid.*, III, 510.
35. *ibid.*, V, 370.
36. Ammianus Marcellinus, III, 15, 13; range estimated from distances achieved by Karamajong/Turkhana tribesmen (see p. 81).
37. Heron, *Bel W*, 75; Heron of Alexandria lived in the second half of the second century AD and was the author of an authoritative manual on contemporary artillery weapons.
38. Vegetius, *op. cit.*, II, p. 15.
39. E.W. Marsden, *Greek and Roman Artillery, Historical Development* (Oxford, 1969), I, p. 15.
40. *ibid.*, and also in his work *Greek and Roman Artillery: Technical Treatises*, Marsden provides considerable detail. See also Payne-Gallwey, *op. cit.*, fn 22, above.
41. *ibid.*, III, p. 83.
42. Payne-Gallwey, *op. cit.*, Pt III, p. 25, fn 1.
43. Philon, *Bel*, 76, as in Marsden, *op. cit.*, IV, p. 94.
44. Payne-Gallwey, *The Crossbow* (London, 2nd edn., 1958), IX, pp. 44–5.
45. Ammianus Marcellinus, xxiii, 4, 4–7.
46. Polybius records that in 250 BC Rhodes sent to Sinope ¾ ton of women's hair for her war with Mithridates (iv, 56, 3); and that in 225 Seleucus made a gift of several tons of hair to Rhodes (v, 89, 9).
47. Josephus, *op. cit.*, II, 548.
48. *ibid*, V, vi.
49. Vegetius, *op. cit.*, II, p. 15.
50. Caesar, *de Bello Gallico*, II, 8.
51. Tacitus, *Annals*, I, 56.
52. Josephus, *op. cit.*, III, 112, and V, 36; also Arrian, *Alani*, 5.
53. Vegetius, *op. cit.*, II, p.15.
54. This is discussed in Marsden, *op. cit.*, pp. 8, 192–4.

Chapter Six

1. Dio's account of a speech by Mark Antony upon the murder of Julius Caesar: IV, xliv, 43.
2. Dio, *Roman History*, 28, 3.
3. Later, in the year 298, after a great victory by Galerius, a peace agreement with Persia advanced the border yet further, along a line from Singara, east of Sura, across the Tigris, to a position south of Lake Van; here it turned sharply westwards, once again to join the old boundary. This new arrangement benefited Rome for it rendered her eastern borders more secure but it was an agreement which lasted barely thirty years.
4. The Straits of Gibraltar.
5. Josephus, *The Jewish War*, II, 367.
6. Strabo, *The Geography*, II, 5, 1, 6–7.
7. Tacitus, *Annals*, IV, 4.
8. Strabo, *op. cit.*, 5, 4, 8–9.
9. Dio, *op. cit.*, VII, LX, 21.
10. Strabo, *op. cit.*, 4, 1, 3–4.
11. Tacitus, *The Histories*, 3, 43.
12. Raymond Chevallier, *Roman Roads* (London, 1989), I, p. 61.
13. Lionel Casson, *Ships and Seamanship in the Ancient World* (Princeton, 1971), pp. 2, 29, Appx 2.
14. Chevallier, *op. cit.*, III, p. 162.
15. Strabo, *op. cit.*, 4, 1, 2.
16. *ibid.*, 4, 1, 14.
17. Chevallier, *op. cit.*, pp. 3, 169.
18. Strabo, *op. cit.*, 4, 3, 2.
19. Tacitus, *Annals*, XIII, 53.
20. *ibid.*, XI, 20.
21. Caesar, *de Bello Gallico*, I, 39.
22. Chevallier, *op. cit.*, pp. 3, 162.
23. Sean McGrail, 'Cross Channel Seamanship and Navigation in the Late First Millenium BC', *Oxford Journal of Archaeology*, pp. ii, 3 (1983).
24. *Wiltshire Archaeological Magazine*, XLVII.
25. John Wacher, *Roman Britain* (London, 1978), pp. 6, 179.
26. Tacitus, *Annals*, XII, 30.
27. John Peddie, *Invasion, The Roman Conquest of Britain* (Gloucester, 1987), pp. 4, 66–88.
28. Josephus, *op. cit.*, II, 367.
29. Strabo, *op. cit.*, XII, 8, 11.
30. Casson, *op. cit.*, pp. 7, 141.
31. *ibid.*, fn 1.
32. *ibid.*, fn 2.
33. Caesar, *de Bello Gallico*, III, 7–16.
34. Dio, *op. cit.*, 50, 32.
35. Casson, *op. cit.*, 13, p. 313.
36. Ammianus Marcellinus, XXIV, 1, 4.
37. *ibid.*, XXIV, 7.
38. Caesar, *de Bello Gallico*, IV, 26.
39. Casson, *op. cit.*, pp. 11, 248, fn 91.
40. Dio, 49, 17, 2.
41. Caesar, *de Bello Gallico*, IV, 23.
42. Diodorus, 20, 51, 1.
43. *ibid.*, 13, 46, 3.
44. A. Dain, *Naumachia* (Paris, 1943), p. 30.
45. Caesar, *War Commentaries, The Civil War*, ed. John Warrington (London, 1965), II, p. 229.
46. John Paul Adams, *Logistics of the Roman Army* (Yale, 1976), II, p. 142.
47. Caesar, *de Bello Gallico*, IV, 24.
48. Caesar, *de Bello Africo*, 20.
49. J.F.C. Fuller, *Julius Caesar: Man, Soldier and Tyrant* (London, 1965), p. 316.
50. See Introduction and Chapter 3, pp. 43, 49.
51. Ammianus Marcellinus, XXIII, 3, 9.

Chapter Seven

1. Michael, Calvert, *Prisoners of Hope* (London, 1952), p. 282; a *machan*, an Urdu word, is a tree-top platform used by big game hunters.
2. For further reading see F.M. Stenton, *Anglo-Saxon England* (Oxford, 1971); J. Peddie, *Alfred the Good Soldier* (Bath, 1989).
3. Calvert, *op. cit.*
4. Caesar, *de Bello Gallico*, vii, 14–31; vii, 68–90.
5. *ibid.*, vii, 4.
6. *ibid.*
7. *ibid.*, pp. vii, 14.
8. Frontinus, *Stratagems*, III, vii, 4.
9. Caesar, *de Bello Gallico*, vii, 23.
10. Appian, *Iberica*, p. 90.
11. Caesar's War Commentaries, *de Bello Civili*, iii, 269.
12. *ibid.*, p. 270.
13. Caesar, *de Bello Gallico*, vii, 22.
14. *ibid.*, vii, 18.
15. *ibid.*, vii, 19.
16. Livy, *The War with Hannibal* (Penguin Classics, Harmondsworth, 1965), xxiv, 34.
17. Jacques Boudet (ed.), *The Ancient Art of Warfare* (London, 1966), Vol. I, Chart 8, p. 128.
18. Josephus, *The Jewish War*, III, 204–32.
19. *ibid.*, v, 162.
20. Aeneas the Tactician, *The Defence of*

Fortified Positions, xxxv.

21. *ibid.*, xxxiii.
22. *ibid.*, xxxiv.
23. Josephus, *op. cit.*, v, 472.
24. Livy, *op. cit.*, xxvi, 45.
25. *ibid.*, xxxviii, 7.
26. Polybius, The *Histories*, ix, 41.
27. *ibid.*
28. See *Illustrated London News*, Archaeological Section 2150, 21 September 1963, p. 425: a brief account of the Leeds–London Universities Expedition to Motya in collaboration with the Mission Archaeologique Française.
29. Arrian, *Anabasis of Alexander I*, I, pp. xx–xxiii.
30. Livy, *op. cit.*, xxiv, 34.
31. *ibid.*
32. *ibid.*, xxvi, 47.

33. As in E.W. Marsden, *Greek and Roman Artillery, Historical Development* (Oxford, 1969), pp. vi, 117, which, for a full account of the working on ancient artillery weaponry, should be read with his *Greek and Roman Artillery, Technical Treatises* (Oxford, 1971).
34. *ibid.*
35. Josephus, *op. cit.*, iii, 158–83.
36. *ibid.*, v, 279.
37. 2 Chronicles 32:30: 'This same Hezekiah also stopped the upper watercourse of Gihon, and brought it straight down to the west side of the City of David.'
38. Josephus. *op. cit.*, v, 25.
39. *ibid.*, v, 302.
40. See Appendix 4 for chronological sequence of events.
41. Josephus, *op. cit.*, v, 472.

Bibliography

PRIMARY SOURCES

Aeneas, *The defence of fortified positions*
Ammianus Marcellinus, *Rerum gestarum libri*
Appian, *Iberica*
Arrian, *Anabasis of Alexander I*
Asclepiodotus, *Tactics*
Caesar, *de Bello Alexandrino*
——, *de Bello Africo*
——, *de Bello Civili*
——, *de Bello Gallico*
——, *de Bello Hispaniensi*
Cassius Dio, *Historia Romana*
Cicero, *Ad Familiares*
Diodorus, *Bibliotheca historica*
Frontinus, *Strategemata*
Heron, *Belopoeica*
Homer, *Iliad*
Josephus, *The War of the Jews*
Livy, *History of Rome*
Onasander, *The General*
Philon, *Belopoecia*
Pliny the Younger, *Epistulae*
Plutarch, *Julius*
Polybius, *Rise of the Roman Empire*
Procopius, *Anecdota*
Strabo, *The Geography*
Suetonius, *Augustus*
——, *Julius Caesar*
——, *Claudius*
Tacitus, *Agricola*
——, *Annals*
——, *Germania*
——, *Histories*
Vegetius, *Epitoma rei militaris*
Velleius Paterculus, *Early Roman History*

SECONDARY SOURCES

Adams, John Paul. *Logistics of the Roman Army*, Yale, 1976.

Boudet, Jacques. *The Ancient Art of Warfare*, London, 1966.

Boyle, Patrick and J. Musgrave-Wood. *Jungle, Jungle, Little Chindit*, London, 1944.

Breeze, David. 'The Logistics of Agricola's Final Campaign', *Talanta*, 16–19, 1987–8.

Casson, Lionel. *Ships and Seamanship in the Ancient World*, Princeton, 1971.

Calvert, Michael. *Prisoners of Hope*, London, 1952.

Chevalier, Raymond. *Roman Roads*, London, 1989.

Von Clausevitz. *Principles of War* (repr.), London, 1943.

Eady, Major H.G. 'The Tank', *United Services Journal*, 1926.

Dain, A. *Naumachia*, Paris, 1943.

Davies, R.W. 'The Roman Military Medical Service', *Sonderdruck aus dem Saalburg-Jahrbuch*, 1970.

Dodge, Colonel T. 'Caesar', *Great Captains Series*, Boston, Mass., 1892.

Engels, Donald W. *Alexander the Great and the Logistics of the Macedonian Army*, Berkeley, 1978.

Farmer, Dr H.G. *The Rise and Development of Military Music*, London, 1912.

Feller, Arthur. *The Fall of the Roman Empire: The Military Explanation*, London, 1986.

Frere, S.S. and J.K.S. St Joseph. *Roman Britain from the Air*, Cambridge, 1983.

Fuller, Major General J.F.C. *Julius Caesar, Man, Soldier and Tyrant*, London, 1965.

Froude, J.A. *Caesar, A Sketch*, London, 1886.

Gudea, Nicolae. 'The Defensive System of Roman Dacia', *Britannia*, vol. 10, 1979.

Hardy, Robert. *Longbow*, Cambridge, 1976.

Harris, W.V. *War and Imperalism in Republican Rome, 327–70 BC*, Oxford, 1979.

Holmes, Dr T. Rice. *Caesar's Conquest of Gaul*, Oxford, 1911.

Hyland, Ann. *Equus, The Horse in the Roman World*, London, 1990.

Keppie, L. *The Making of the Roman Army*, London, 1984.

Last, Professor H. *Cambridge Ancient History*, Cambridge, 1932.

Lewis and Short, *Freund's Latin Dictionary*, Oxford, 1879.

McGrail, Sean. 'Cross Channel Seamanship and Navigation in the late First Millennium BC', *Oxford Journal of Archaeology*, 1983.

Marsden, E.W. *Greek and Roman Artillery, Historical Development*, Oxford, 1969.

——. *Greek and Roman Artillery, Technical Treatises*, Oxford, 1971.

Mommsen, Theodor. *The History of Rome*, London, 1911.

Montgomery, Field Marshal Viscount. *Concise History of Warfare*, London, 1968.

Pakenham, T. *The Boer War*, London, 1979.

Payne-Gallwey, Sir Ralph. *The Projectile Throwing Engines of the Ancients*, London, 1907.

——. *The Crossbow*, 2nd edn, London, 1958.

Peddie, John. *Alfred, The Good Soldier*, Bath, 1989.

——. *Invasion, The Roman Conquest of Britain*, Gloucester, 1987.

Richmond, Sir Ian. *Trajan's Army on Trajan's Column*, British School at Rome, London, 1982.

Richmond, I.A. and J. McIntyre. 'Tents of the Roman Army and Leather from Birdowald', *Transactions of the Cumberland and Westmoreland Archaeological Society*, 1934.

Sachs, Curt. *The History of Musical Instruments*, London, 1940.

Spiedel, M.P. *Eagle Bearer and Trumpeter*, Rheinisches Landesmuseum, Bonn, 1967.

Slim, Field Marshal Sir William. *Defeat into Victory*, London, 1956.

Stenton, F.M. *Anglo-Saxon England*, Oxford, 1971.

Tarn, W.W. *Hellenistic Military and Naval Developments*, Cambridge, 1930.

Wacher, John. *Roman Britain*, London, 1978.

Webster, Graham. *The Roman Imperial Army*, 2nd edn, London, 1979.

Wilmot, Chester. *The Struggle for Europe*, London, 1952.

Index